AI AND BLACK PEOPLE

AI AND BLACK PEOPLE

OBI M HOLLY

CONTENTS

ABOUT THE AUTHOR

This book is dedicated to all Black innovators on the brink of helping us reclaim our rightful place in the world. May the spirits of our ancestors open your divine doors, allowing you to usher in a new era of light and love.

Hetep!

ACKNOWLEDGEMENT

I would like to acknowledge the village that birthed and raised me—the African ancestors and living geniuses who paved the way with their sweat and blood so that I could stand here today and spread the good word. I stand on far too many shoulders to name them all, but I must express my deepest gratitude to my grandmother, Mama Asante Ayetoro, for insisting that my name be Obi. This Nigerian Igbo name, meaning "heart," has been a constant reminder of my African heritage and my duty to connect with and uplift my people.

I want to acknowledge my mother, Ava Hancock (Mut Saukhet Menut), for always standing by my dreams, even in difficult times, and my father, Thomas B. Holly, for keeping my ears to the streets and for the deep spirituality, love, positivity, and courage that you have shown our entire family.

I also want to thank Ur Aua Hehi Metu Ra Enkhait of the Ausar Auset Society for naming me Sauti Tut Kham, meaning "one who strives to be as the ancient sage." This name has guided me to stay focused on my spiritual upliftment and always to strive to embody the divine attributes of our ancestors.

To Baba Agyei Akoto (Fly In Peace), thank you for being the image of a true Black man and father, working daily for the advancement of African people. To Shekhem Ur Shekhem Ra Un Nefer Amen I, King of Kings of the Ausar Auset Society, I express my deepest appreciation for helping to reawaken my spirit and that of the masses with the knowledge of our birthright to ancient Kamit and how to make it manifest once again.

I extend my gratitude to the DC Council of Elders, the Akan, Yoruba, and all other African traditions, as well as the many Black organizations working tirelessly for the prosperity of our people worldwide.

To my children—Amsara Sekhita Mai Holly, Taia Maati Khum Holly, Maasatimbala Sanefer Miling Holly, and Ariana Shring Holly—thank you for your patience as I pursued my visions. And last, but certainly not least, to my Queen, Amenika The Great / Nana Adowa Asamaniwa I (Amenika Shepsut - Carla Holly), thank you for your wisdom, creativity, love, and always supporting me and standing strong as I led us down so many paths in pursuit of our greatest destiny. I love you!

I would like to thank Amsara Sekhita Mai Holly for editing and creating the AI-generated imagery, and Shekhem Tepraim Saa for providing valuable insight during the development of this book.

Thank you all, and may God bless you!

Amen Ra - Ptah! WE ARE BACK!

CHAPTER 1

Introduction

When I was a kid, I always knew computers were the future. At 12 years old, way back in 1982 (besides playing in local go-go bands), I started (along with a concerned parent) the first computer club at my independently owned and operated Black private school (Nation House Positive Action Center - Watoto Shule). Nobody in the hood, and I mean nobody, had even seen a computer back in those days (unless their parents worked for Xerox). Needless to say, this was a pretty big deal.

They purchased a few Commodore 64 computers, and we were off to the races. We learned how to use computers by making simple games using the "Basic" computer language. It was fun, but the games were nowhere as dope as Space Invaders and Pac-Man, which came to market a few years earlier on the Atari gaming console. Public arcades and home arcade games were all the rage, and our Commodore 64s teased us into believing we had what it took to be the next great game creators.

Before that, in the 70's, I was a serious fan of the film Star Wars. In part because my mom took me out of a neighbourhood school and put me in an all-white elementary school for a few years. These white kids were obsessed with Star Wars films and would camp out for days to get tickets at a movie theatre a few minutes from our school. My birth name is Obi, so ever since Star Wars was released, everyone has called me Obi-Wan, Ben Kenobi, Ben, Obi-Wan Kenobi, or Obi-1.

So, as my wife says… "I got it honestly." I, Obi-1, AKA Obisanichiban (that's the Japanese version for Obi-1– given to me by Mami Shirikawa, the former host for Black Entertainment Television's *Rap City Japan*, a show I used to produce in the '90s…but that's a different story!) have always been curious about the future and what trends will take us to the next level in human spiritual and technological evolution. We may soon be approaching an era where our technology may even surpass and out-smart the android intellect of R2D2 and C3P0—sorry, Star Wars.

As a Black person, I don't want our people left out of the creative side of the next wave of human technological innovation. Since these new technologies are and will continue to affect our lives, we should take an active role in creating the most positive and powerful effects. Let us not forget a cautionary statement from Dr. Martin Luther King Junior..."*Modern man has brought this whole world to an awe-inspiring threshold of the future. He has reached new and astonishing peaks of scientific success. He has produced machines that think and instruments that peer into the unfathomable ranges of interstellar space. He has built gigantic bridges to span the seas and gargantuan buildings to kiss the skies. His airplanes and spaceships have dwarfed distance, placed time in chains, and carved highways through the stratosphere. This is a dazzling picture of modern man's scientific and technological progress. Yet, despite these spectacular strides in science and technology, and still unlimited ones to come, something basic is missing. There is a sort of poverty of the spirit which stands in glaring contrast to our scientific and technological abundance.*"

When Dr. King spoke these words, Apple, Microsoft, and Google didn't exist, and NASA had less computing power in their entire headquarters than the average 12-year-old has on their cell phone in their back pocket. Marinate on that for a minute.

CHAPTER 2

Importance of AI for black People

"Armageddon, it's been in effect...go get a late pass! Step!" These are the words spoken by the legendary rap group Public Enemy on their groundbreaking hip-hop masterpiece, *It Takes A Nation of Millions To Hold Us Back*! This statement to me warned the audience that if you are just figuring out how dire the situation is, then you are perhaps too late. In essence...it's time to wake up and catch up!

Artificial Intelligence (AI) is by no means a recent development; early research in this field dates back to at least the 1950s. Unbeknownst to many, you've been incorporating this technology into various aspects of your life for years. Its seamless functionality and commonplace applications might have escaped your notice, but don't sleep—it has been there, growing, learning, and consistently improving. While you were effortlessly sending emails, handling tax returns, recording music, creating blueprints, debugging code, enjoying advanced refrigerator features, utilizing photo face recognition tools, and managing daily planners, something significant was quietly evolving in the background.

Now, with the remarkable technological advancements in Generative AI and other tools (such as high-powered drones and robots capable of intricate movements), which our collective data unwittingly contributed to, you are starting to realize that a transformative shift is underway. This transformation is occurring rapidly, unfolding right before our eyes. *"Armageddon, it's been in effect...go get a late pass! Step!"*

It has long been theorized that computing technology, the foundation for artificial intelligence (AI), follows the principles of "Moore's Law." This law dictates that computing power experiences exponential growth, with capabilities and performance doubling over a short period, typically every 18 months to 2 years, as transistors become smaller and more densely packed on integrated computing circuits. Consequently, the speed and capabilities of our computers and devices have been and are expected to continue to advance while becoming more affordable.

With numerous technological advances since the inception of "Moore's Law", many tech insiders have claimed that Moore's Law is dead. The fact is that contemporary technology utilizes various types of new tools, a variety of transistors, and other variables, which enable new tech to surpass the projections of this law.

The transformative changes in Artificial Intelligence (AI) that may seem to have caught us by surprise have been in progress for years. The recent acknowledgment of its advancement is primarily attributed to the rapid expansion of Machine Learning (Natural Language Processing, Computer Vision, etc). In this process, machines go beyond being programmed solely to perform a function; they are designed to correct past errors, adapt to evolving situations, identify patterns, and essentially "learn" to improve performance across various tasks. This surge in capabilities is complemented by the emergence of innovative Generative AI technologies that process human language and create a tailor-made response. Natural language processing algorithms like ChatGPT, SORA/SORA2, Bard, Mid Journey, DeepSeek, and others generate content based on human user prompts (a prompt is the instruction or input you give to a generative AI to tell it what you want it to create or do). The fascination and utility of the amazing results (artistic, written documents, presentations, etc) generated by the AI have contributed to a dynamic and engaging social media frenzy surrounding the new technology as well as major shifts in multiple industries. As AI continues to expand its influence on society, it becomes the responsibility of all communities, especially Black people, to stay informed and actively participate in its development and deployment.

While Black folks have always been on the cutting edge of technological innovation since the beginning of civilization, Black people have also been historically underrepresented (and /or not given credit for their innovation and inventions) in the industrial and technology industry in the United States (and Western society). According to a 2023 Report, Black people make up 7% of the 9.2 million tech industry workforce, while white people make up 63%, Asian Americans make up 20%, and Hispanic or Latino people make up 8%. An analysis of Fortune 500 executives found that only 3% identify as Black. This lack of representation has often resulted in AI systems that are not inclusive and will (and are) perpetuating bias and detrimental existing inequalities through the new platforms and applications. Thus, I created AI and Black People.

This book will provide quick information and resources to help Black people understand and engage with AI meaningfully. The book will aim to:

1. Raise awareness about the impact of AI on Black communities and how it is likely to impact us.
2. Establish the fact that black people have always been pioneers in technology.
3. Empower Black people to become involved in the positive development and deployment of AI systems.
4. Address the issue of bias in AI algorithms and provide strategies for detecting and preventing bias in AI systems.
5. Encourage Black people to consider careers in AI development.
6. To remind the Black community that despite the excitement surrounding the new innovation associated with AI, nothing has surpassed the genuine development of the mind and spirit. In essence, AI is merely a tool we should utilize to channel its power for constructive transformation. The resolutions to our profound questions and challenges will, in the end, hinge on the elevation and broadening of both our individual and collective consciousness. Ultimately, it is through the strength of our unified minds and spirits that we can pave the way for a brighter future for everyone.

Important Note: Besides using the abbreviated term AI (Artificial Intelligence), two other key terms often repeat themselves in this book. I would like to clarify what I am talking about when using these terms.

Technology:

In the context of this book, the term "technology" refers to the practical application of scientific knowledge, particularly in industrial settings. It encompasses the development of machinery and devices derived from scientific principles. This field of knowledge encompasses engineering and applied sciences.

It's crucial to recognize that technology extends beyond the realm of physical mechanics and machinery. Across various cultures worldwide, there exists a multitude that heavily relies on highly advanced "spiritual technology" and rituals within their societies, often with minimal reliance on traditional "high-tech" mechanical implementations (We will discuss this concept later).

Algorithm:

An algorithm is a step-by-step set of instructions your computer or applications follow to solve a problem or complete a task. Algorithms act as an exact list of instructions that conduct specified actions in either hardware- or software-based routines. Humans have their own step-by-step instructions for making a cup of coffee or cooking rice. The algorithm in your software or computing device serves as a recipe or game plan for your applications to follow to produce a desired result. For instance, when utilizing your phone for navigation, the algorithm embedded in your map application is programmed to follow certain steps to identify the route. Furthermore, it is also programmed to "learn" as you travel by gathering diverse forms of additional information, thereby presenting alternate, quicker routes. Algorithms are widely used throughout all areas of AI, IT, and computing.

Now that we have established the basics...Let's move forward.

Black To The Point:

Let's become aware of AI. Let's examine the good and bad possibilities. Let's discover how we can benefit from this exciting, rapidly evolving technology.

CHAPTER 3

What Is AI?

Artificial Intelligence (AI) refers to the simulation of human intelligence in machines designed to perform tasks that typically require human intelligence, such as understanding language, recognizing patterns, making decisions, etc...

These simulations of human intelligence can be used in a multiplicity of ways and placed in machines or remote locations such as the Cloud or internet, and used to operate a variety of devices, including but not limited to these common tools listed below:

- Phones
- Security Systems
- Home Appliances
- Televisions
- Virtual and Augmented Reality Systems
- Automobiles
- Computers
- Robotics
- Cameras
- Websites

- Health Care Devices
- Recording equipment
- Human Translation Devices...Even Animal Translation Devices (maybe)
- Art Applications
- Stock Market Applications
- Writing Applications
- Virtual Reality Worlds
- Nano Technology

Some form of AI is commonly present in virtually any domain where computer systems and algorithms find application. A recent advertisement caught my attention, showcasing a hamburger place in LA that operates entirely through AI-driven French fry and burger-making machines. While this tech is definitely "cool," the innovation implies a reduced demand for human labor in these types of minimum-wage jobs. That reduced demand can, will, and has affected the job market, making some jobs obsolete and opening the door for new innovations and innovators. Additionally, during a recent hotel stay, I experienced an AI-driven robot on wheels delivering extra towels and pillows to my room. The machine sported a digital smile and emitted a melodic whistle sound as it departed, almost like saying. "Goodbye!" Where was the cleaning lady? Who do I give this tip to? I don't know! In LA, AI-powered delivery robots on wheels are rolling around everywhere from Hollywood to Long Beach. These robots use a combination of cameras, sensors, and AI to navigate sidewalks and make food and grocery deliveries. The possibilities for AI applications are endless.

AI systems can be classified into two types:

Narrow or Weak AI:

Narrow AI, also referred to as specialized AI, is meticulously crafted to excel in a single task or a limited range of tasks while consistently improving its performance. Its primary objective is to identify automated solutions for problems and inconveniences or enhance existing systems for optimal functionality. Currently, the majority of Artificial Intelligence (including Generative AI like ChatGPT and others) falls within the realm of Narrow AI. These AI systems are typically software-based, designed to automate tasks traditionally undertaken by humans, often surpassing or aspiring to surpass human capabilities in terms of efficiency and endurance. It is important to keep in mind, however, that Narrow AI, no matter how amazing, cannot think, reason, or adapt like a human.

Examples of Narrow AI:

- Siri or Alexa

- Spotify recommendations
- Google maps
- Chatbots on websites
- Netflix Suggestions
- Facial recognition technology
- Smartphone apps that provide accurate weather predictions
- Self-driving cars, such as those developed by Waymo / Google, and others...

Side Note: I worked for Google Street View a few years ago and witnessed people returning from "test rides" in self-driving cars at the Google offices in Mountain View, California. Passengers, on the driver's side, would often be engrossed in their smartphones, trusting that the car's AI system would reach its destination safely. I remember thinking, "Hmm, I wonder how that's going to work out?"

While it's still a work in progress, Waymo, the autonomous driving technology company that was formerly known as the Google Self-Driving Car Project, seems to be working it out pretty well with its AI rideshare service in the streets of LA and other select cities. The Black CO-CEO of Waymo, Tekedra Mawakana, appears to be laser-focused on ensuring the safety of the massive AI fleet. It's pretty wild to see an autonomous "driverless" car pull up next to you at a red light.

General or Strong AI:

General or Strong AI possesses the capability to execute a variety of tasks akin to those performed by human beings. Some designate it as "The True AI" because it signifies the next phase in advancing machine intelligence to a more comprehensive level. Instead of being confined to a singular task like Narrow AI, the goal is to empower machines to understand and reason at a broader scope, mimicking human cognitive abilities. General or Strong AI can include two types of AI:

- Artificial General Intelligence (AGI): A self-aware system with consciousness that can solve problems and plan for the future. The development of AGI would represent a major milestone in AI research, as it would involve creating machines that can perform any intellectual task that a human can do, including reasoning, problem-solving, learning from experience, and understanding complex concepts.
- Artificial Super Intelligence (ASI): A hypothetical form of AI that surpasses human capabilities in every aspect, including problem-solving, creativity, and emotional understanding. ASI is purely theoretical and raises many ethical and existential concerns.

The ultimate aspiration is to create machines with the capacity to think in a general sense, making decisions based on acquired knowledge rather than relying solely on predetermined training. These machines would possess the ability to assess their training and determine if there are more suitable courses of action to pursue. The desired outcome is independent learning from experience, mirroring how humans learn and reason. Imagine, for example, a robot that could learn new skills on its own, like cooking, playing music, or coding while holding a conversation, doing your taxes, painting a picture, and building a house - all without needing to be programmed. Currently, all real-world AI is Narrow, but many individuals are diligently working to create real General AI.

Some people are even seeking Artificial Immortality. A few Church organizations have sprung up that seek to one day use AI technology to transform human consciousness into a database, allowing people to actually "live forever" digitally. While this idea may be far off, it is in fact possible to input personal photos, videos, written documents, audio recordings, and other content about yourself into a computer. That data can be combined with your image and voice to create a digital clone (and full robot body) of you that friends and relatives can interact with long after you die. It's crucial to note that a digital clone at this point is not "sentient" or "self-aware." It functions as a tool following its programming, even if that programming allows it to "learn" and appear life-like. Thus, despite AI's accuracy and ability to mimic some human intelligence and blow your mind... it still lacks soul... And SOUL is very important to Black Folk!

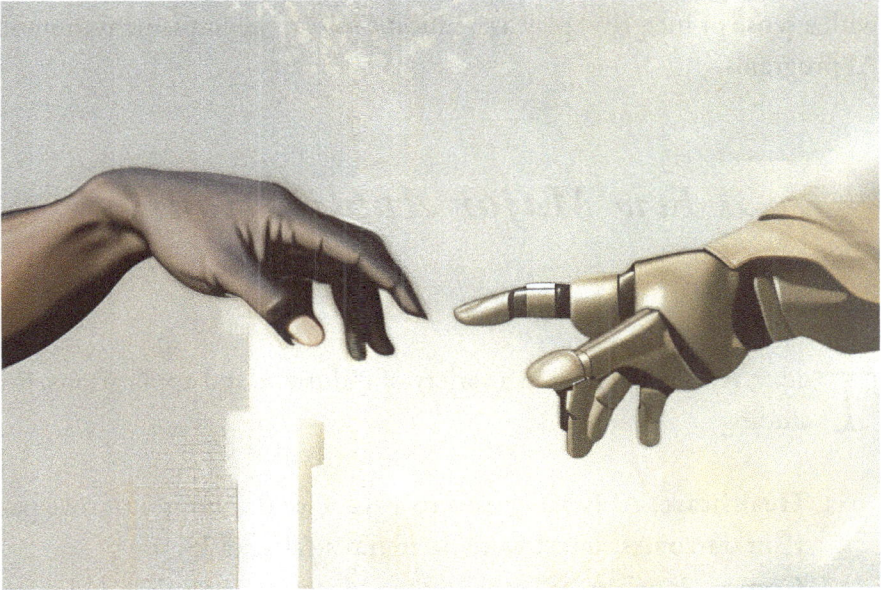

AI systems operate by utilizing algorithms and statistical models to analyze data and make predictions or decisions. These algorithms are designed based on rules and instructions provided by developers. As stated earlier, many AI systems can learn from the information and data they encounter while performing tasks, improving performance over time.

When it comes to addressing issues that specifically impact marginalized groups, such as Black people, the concept of "diversity" in AI development becomes crucial. The effectiveness and accuracy of how an AI system tackles certain problems affecting Black individuals, for example, will largely depend on the individuals programming the AI and the data (information, internet, laws, history, etc.) the system is trained on. Therefore, it is of utmost importance that Black people (especially those with a sense of integrity) play a significant role in the implementation of AI programs.

A Few Major Applications:

Today, AI is being used in a variety of industries and applications, including:

1. **Healthcare:** AI is being used to assist with diagnosis, improve patient outcomes, and streamline administrative tasks.
2. **Companionship:** Whether due to loneliness, a need or desire for some form of social interaction (or artificial interaction), or simply as a form of entertainment, AI companions are used for emotional and mental health support, elderly care, virtual games, and romantic simulations, to name a few.
3. **Robo Assistance:** AI assistants provide personalized, voice-activated support for tasks like scheduling, information retrieval, and home automation. They learn from user interactions to offer tailored recommendations, improve over time, and adapt to individual preferences and habits.
4. **Agriculture:** AI offers innovative solutions to traditional farming challenges. By examining results, machine learning, and automation, AI applications can enhance productivity, optimize resource usage, and promote sustainable farming practices.
5. **Robo Advisors:** AI advisors provide expert-level recommendations and decision support by analyzing vast amounts of data,

identifying patterns, and predicting outcomes in areas like finance, healthcare, and business strategy. They offer insights, simulations, and scenario analyses based on real-time and historical data.

6. **Social Media:** Content personalization, Image and video analysis, Sentiment analysis, Chatbots, Spam, and fake news detection, deep fakes, and social media content generation.

7. **Finance:** AI is being used to help with fraud detection, risk management, customer service, stock market predictions, budget creation, etc...

8. **Transportation:** AI is being used to improve traffic flow, reduce accidents, and optimize supply chain management. AI-driven vehicles and drones.

9. **Retail:** AI is being used to personalize customer experiences, improve product recommendations, and enhance supply chain management.

10. **Manufacturing:** AI is being used to optimize production processes, reduce waste, and improve quality control.

11. **The Military:** Robots are being created that can participate in war. While this is nothing new, AI algorithms are being programmed into machines that can do everything from fly, run, jump, kick, punch, and shoot to kill!

12. **Generative AI / Arts and Entertainment:** New applications create complicated pictures and images in seconds. Other applications help authors write entire scripts and books, while others produce and edit music and videos.

13. **Disinformation:** AI can generate and spread false content through deepfakes and social media bots, making it harder to distinguish truth from falsehood. Conversely, AI is also used to detect and combat disinformation by identifying fake news, analyzing

content patterns, and flagging misleading information for removal or correction.

The reality is that AI technology is impacting everyone on the planet in some way or another. If it's not currently employed in your industry, it's likely to be integrated soon. The phone in your pocket, not to mention social media applications, is replete with AI algorithms that operate continuously, even as we sleep.

Your cell phone, for example, has long used various algorithms:

- **Personalized recommendations:** AI algorithms that analyze your behavior and preferences to provide personalized recommendations for apps, content, and services.

- **Image and video recognition:** This AI is used in smartphones to recognize faces, objects, and scenes in photos and videos, enabling features such as portrait mode and automatically tagging people in images.
- **Speech recognition and virtual assistants:** AI-powered speech recognition enables users to interact with their devices using natural language. Virtual assistants such as Alexa, Siri, and Google Assistant use AI to respond to voice commands and provide information and assistance.
- **Predictive text and autocomplete:** AI algorithms are used to predict the words and phrases a user will likely type next, enabling more efficient and accurate typing on small screens.
- **Battery optimization:** AI algorithms can analyze a user's device usage and adjust the power settings accordingly, helping to extend battery life.
- **Augmented Reality (AR):** AR apps use AI to understand the real-world environment and place digital objects and information in a user's view of the world.

Black To The Point:

AI is here to stay (unless all technology is suddenly destroyed). It does many jobs humans used to do only faster, "smarter," and "better." It's time to diversify your working skill set. Your current job may become obsolete. It's time to either level up and embrace AI, change careers, or risk being left behind. My bad.

CHAPTER 4

IMHOTEP

A Brief History of African People and Technology

B lack people have a rich history in the development, use, and implementation of technology. In fact, we were the pioneers, the first on Earth to extensively harness technology in ancient times. The utilization of agriculture for cultivating crops, the mastery of architecture in constructing buildings, cities, and temples, proficiency in high mathematics, astronomy, the creation of writing systems, the science of drumming communication, shipbuilding, ocean navigation, and more – all these innovations were conceived and brought to fruition by Black people in ancient Africa (and beyond) long before these systems disseminated to others worldwide. The assertion that Black people are the originators of advanced technology is not rooted in Afrocentric thinking; it is simply a reflection of scientific truth.

Allow me to stop down for one minute to let you know that this book will not engage in the debate over the "Blackness" of the indigenous originators, founders, and maintainers of ancient African Egyptian civilization. There is no controversy as far as I'm concerned. Everyone in ancient times, according to their own accounts, knew who the Ancient Egyptians were. Besides the countless paintings and stone statues, Herodotus, the Greek historian, the Romans and others all put the indigenous ancient Egyptians / Kamau into the category of what you call today a "Black person" (we can debate at another time if people who call themselves black should continue to use this term to describe themselves. For the sake of this book and universal understanding, I'm using the word Black!). "Kamit" means the Black land or land of Black people, and the term Kamau...what they called themselves literally means Black people (who, of course, come in multiple complexions). The only debate exists when misinformed or racists (be they historians, film or documentary makers, writers, politicians, geneticists, tour guides or others) attempt to cover up the truth (even through DNA scams) to fulfill their twisted and false agendas, controlling the narrative and confusing the world often for profit (Please Look up the book "Black Pharaoh: African DNA and Anthropology of The Ancient Egyptians" by Enensa A. M. Amen). We also must keep in mind that modern Egypt, unlike ancient Kamit, is a country with way over 2000 plus years of imperial colonization and mixing by outside non "Black" peoples and cultures (Hyksos, Assyrians, Persians, Greeks, Romans, French, English, The Ottoman Empire, The Rashidun Caliphate and others (Look at what just 400 years of European colonization of the area now known as the United States of America has done to the indigenous population). This has had a profound effect on the modern population of Egypt, creating a totally new demographic as compared to ancient Black Egypt / Kamit. The indigenous originators and maintainers of ancient Egyptian society and culture were Black people. Black people all over the world are awakening to this fact, and there is nothing anyone can do to stop it! Period! Moving on!

Imhotep:

While there have been numerous African geniuses who made significant contributions to technology throughout history, one standout figure is Imhotep, who gained particular acclaim, especially among the Ancient Greeks. Despite living thousands of years before Alexander the Great, Imhotep was revered and even deified by the Greeks as a God of medicine and healing. Imhotep, a Black African man, wore many hats, serving as a doctor, mathematician, African priest, architect, and scribe (to name a few) – a true multi-genius.

In his role as the chief advisor for Pharaoh Djoser of the 3rd dynasty, Imhotep is credited with the design and construction of the Step Pyramid in Sakkara, Egypt, marking the first pyramid ever built and perhaps one of the earliest colossal stone structures in Egypt. An intriguing note arises when examining Hollywood's portrayal of Imhotep in the 1999 movie "The Mummy," depicting him as an evil undead monster bent on destroying the world. This irony is compounded by the fact that the name Imhotep literally translates to "he who comes in peace." One can only ponder the repercussions if Hollywood were to depict other revered figures like Mahatma Gandhi, Buddha, or non-African historical spiritual leaders as disrespectfully.

Unfortunately, because of misinformation, many Black people may not fully grasp their connection to some of their historical heroes or the vital role they played in laying the groundwork for civilization and technology. This lack of awareness often leads to a passive response, with masses of people being programmed with lies and even enjoying disrespectful entertainment rather than expressing outrage and taking action.

I   EM   HETEP – Imhetep / Imhotep / AIM HETEP

He Who Comes In Peace

Even today, one can find ancient hieroglyphs of doctors' instruments chiseled into stone temples throughout modern Egypt. These items, along with many other writings, artifacts, and megalithic monuments, solidify the fact that ancient Black Africans had mastery over ancient medicine, mathematics, agriculture, measurement, astronomy, astrology, writing, and all of the other major building blocks of "civilization" and technology.

Imhotep stands as a prominent figure, but he is not the sole multi-genius in ancient Africa or ancient Kamit (known as Egypt in Greek). Imhotep serves as just one template among many geniuses who preceded and succeeded him. This civilization, boasting over 3500-plus years of remarkable technological advancements, left behind evidence of high-tech achievements that still perplex modern scientists. Within this ancient society, numerous geniuses thrived. The spiritual foundation of the ancient Egyptian system empowered individuals to unlock their mental and spiritual potential, fostering mastery in various fields, including but not limited to mechanics and technology. The system of self-mastery cultivated in Kamit positioned it as the global epicenter for formalized higher learning and achievement. This, in turn, heavily influenced the foundation of civilization and technological advancements in Europe and Asia, driven by a mix of appropriation, borrowing, and teaching.

The Greeks, for example, owe a major debt to the ancient Egyptians for their technological advances that, in turn, greatly influenced the Roman civilization (which went on to influence all of Europe). Many Greek scholars were trained in Egypt and took that knowledge back to Greece to help transform their society (seldom giving credit to their Kamitic sources. Please see "Not Out of Greece" by Ra Un Nefer Amen, "Stolen

Legacy" by George G.M. James, "Destruction of Black Civilization" by Chancellor Williams, and many others.)

Though a ton of "non-African" advancements should be considered part of Kamit's stolen legacy due to constant invasion, pillaging, adaptation, and imitation by foreigners, it's essential to acknowledge that Ancient Kamit also generously shared many aspects of its knowledge with the rest of the world.

Ancient Egyptian cultural and technological foundations largely began to the south of Egypt, yet extended in every direction beyond ancient Egypt. To the south of Egypt, upstream along the Nile River in Ancient Nubia / Kush (modern-day Sudan), Africans created numerous temples, cities, and even more pyramids than in Egypt. The roots of ancient Egyptian culture also largely started further south in Africa's interior

(Sudan, Uganda, Ethiopia, etc.) and then moved downstream toward the north, eventually founding ancient Kamit along the Nile River. As Kamit developed, it became a repository of thousands of years of wisdom and knowledge from Africa's interior and continued to exchange goods and ideas with other Black African groups on and off the continent and with other people in the world.

Whether in the north, south, east, or west of Africa, remnants of high technological achievement abound. As far away as Mpumalanga, South Africa, researchers have uncovered Adam's Calendar, potentially the world's oldest man-made structure—an astronomical site often dubbed the "African Stonehenge," dating back perhaps over 75,000 years. Ancient Africans also traversed the continents, sharing their advancements, settling, establishing, teaching, and blending in with numerous civilizations, thus influencing technological growth in far-flung corners of the Earth (though that's a topic for another book).

This section does not aim to delve into all the technological achievements of Black people. Its purpose is simply to instill the idea that Black people have consistently been at the forefront of technological achievement, contrary to what many may have learned in the Western school system or the many other unjust thieves and appropriators of Black culture and historical legacy. Despite current technological disparities in the black diaspora, it's essential not to draw the conclusion that Blacks made little or no contributions; the truth is quite the opposite.

Spiritual Technology

Throughout history, one of the primary focuses for African people has consistently revolved around their level of respect and relationship with nature and the divine. From the very outset, Africans have engaged in highly detailed and advanced spiritual technological practices. Spiritual technology encompasses methodical systems of spiritual cultivation and practices, enabling individuals to transcend behaviors, elevate their minds and spirits, heal themselves and their communities, effect change in their environments and circumstances, and explore communication with the many regions of the "spiritual" (everything is spiritual in the African mind) realm. This involves honoring and communicating with the ancestral world and unifying with "God" and its attributes (the African indigenous concept being vastly different than that of modern Western culture), thus gaining access to spiritual powers and information that many might consider supernatural or inaccessible by physical means.

In numerous African societies and similar communities worldwide, this type of access is viewed simply as natural rather than supernatural. Mastery of the use of herbs, plants, and various elements to invoke both mundane and "spiritual" results, energy healing, diverse breathing techniques, spiritual exercises and energy manipulation, meditation techniques, manipulation of sound vibrations, chants, and frequencies to manifest effects, ancestor communication, the proper naming of individuals, knowledge of planets visible only through telescopes, knowledge of the effects of planetary cycles and how to utilized them and more are seamlessly integrated into daily life and rituals in traditional Black societies globally.

"Modern science" has only recently begun to acknowledge and scratch the surface of what the Dogon people of Mali (for example) have held as traditional norms of spiritual-technological knowledge for thousands of years. From ancient Kamit to the San people of the Kalahari Desert and the Island of Vanuatu, high spiritual technology has been a way of life-science since time immemorial. While some African groups may not have delved into massive mechanical, technological projects (such as pyramid building, which are actually large-scale projects with spiritual purposes), almost all have evolved massive spiritual technology as a foundational aspect of their culture. In contrast to so-called modern society, which increasingly relies on mechanical technology and artificial Intelligence to achieve various goals, our ancient African ancestors (and many today) depended first on "Ancestral Intelligence" and the power of the spirit to attain what cannot be achieved through physical means.

A VERY SHORT LIST OF ANCIENT BLACK AFRICAN TECHNOLOGICAL ADVANCES

The Great Sphinx of Giza (Heru Em Akhet – Horus of The Horizon): This colossal limestone statue, featuring an African face (believed by many to originally be an African woman) and the body of a lion, resides in North Africa on the Giza Plateau in Egypt. It stands as the world's oldest and largest monolithic structure. Estimated to be at least 4,500 to 7,000 years old; however, some interpretations (because of possible evidence of water and flood erosion) suggest an age of 16,000 years or older.

The Pyramids of Giza: The Great Pyramid was constructed for or on behalf of Pharaoh Khufu of the 4th dynasty of Kamit (EGYPT). It held the title of the tallest man-made structure globally for over 3,800 years until the advent of the Lincoln Cathedral in England in 1311. Remarkably, dividing the perimeter of the base of the Great Pyramid by twice its height yields an accurate estimate of the mathematical principle of pi, a feat falsely and traditionally credited to the ancient Greeks (Alexander the Great would not be born for more than 2000 plus years after the construction of the pyramids). This monumental structure is not only a testament to ancient engineering but also a mathematically sound marvel, boasting numerous astonishing attributes. The invitation is extended for all to continue exploring and researching this extraordinary achievement.

King Khufu's Boat: Pharaoh Khufu's ship is an intact full-size solar barque (sailing vessel) from ancient Egypt. It is 43.4 meters (142 ft) long and 5.9 meters (19 ft) wide and is considered the world's oldest intact ship. The ship is over 4,500 years old and can still sail today. The ship is currently on display at the Grand Egyptian Museum in Giza Governorate.

Nabta Megaliths (In Southern Egypt): Older than Stonehenge, huge stone slabs found in Egypt's Sahara Desert are between 6,000 and 7,000 years old and have been confirmed by scientists to be the oldest known astronomical alignment of megaliths in the world.

Adam's Calendar: Perhaps the oldest man-made structure in the world. An astronomical site nicknamed the "African Stonehenge", which may be over 75,000 years old and therefore pre-dates Stonehenge and the pyramids of Giza by tens of thousands of years. Located in Mpumalanga, South Africa, it is a standing stone circle about 30 meters in diameter that appears to be the oldest example of a completely functional, mostly intact megalithic stone calendar in the world.

Kush/Nubian Pyramids: Constructed over several centuries, the Nubian pyramids served as tombs for the kings, queens, and affluent citizens of ancient Napata and Meroë. The initial three sites are situated around Napata in Lower Nubia, near the modern town of Kerima. The first pyramid was erected at the el-Kurru site, housing the tombs of King Kashata and his son Piye, along with Piye's successors Shabaka, Shabataka, and Tanwetamani. Additionally, numerous pyramids were dedicated to renowned warrior queens of ancient Kush. Notably, the Nubians built approximately 255 pyramids, surpassing more than double the count in Ancient Egypt.

Mummification: It is widely acknowledged that the ancient Egyptians in Africa were adept practitioners of mummification for thousands of years. However, the oldest evidence of "advanced mummification" dates back around 5,600 years and was discovered in modern-day Libya. Known as the "Black Mummy of the Green Sahara," this mummy predates the earliest Egyptian mummies by 1000 years. The mummy, depicting a black boy, underwent a process involving the removal of organs, embalming with an organic substance to deter decomposition, and wrapping in cowhide, accompanied by ritualistic items.

The First Paved Road: Dating from the Old Kingdom in ancient Egypt, people transported basalt blocks from an ancient quarry to a wharf on the shores of ancient Lake Moeris. The Lake Moeris Quarry Road, in the Faiyum District of Egypt, is the oldest paved road in the world, and a considerable part of its original pavement is still preserved.

Writing: In addition to the development of various forms of writing in numerous African cultures (from north, south, east and west Africa) independent of the influence of Islam and Christianity, recent discoveries reveal that ancient Egyptian writing, in the early hieroglyphic form, served as "tag-labels" in the ancient pre-dynastic city of Abydos from perhaps 3,800 BCE or before. Subsequently, the Ancient African Egyptians (Kamau) developed at least three major forms of writing, including hieroglyphics (Metu Neter), hieratic, and demotic writing. The use of early hieroglyphics in Abydos laid the groundwork for many writing styles that followed. The "Pyramid Texts" of the ancient African King Unas are considered the oldest religious text in the world. The text inscribed on the walls of the burial chamber in Unas' pyramid date to approximately 2400 to 2300 BCE. The ancient Nubians also created the Meroitic alphabet, which pre-dates Christian and Islamic influence by centuries (I mention this to emphasize the fact that writing is and always has been a cherished ancient African indigenous tradition). Foreign culture did not bring the idea of a book (writing on Egyptian "papyrus" – the origin of the word paper) or reading and writing to Africa or Black people... BLACK PEOPLE GAVE THE WRITTEN WORD TO THE WORLD!

The Ifa Oracle Binary System of 256 Odu: is an ancient divination practice of the Yoruba people of Nigeria, guided by Babalawos (high priests disciples of Orunmila - the Orisha/divinity of wisdom and knowledge). At its core, the system uses either sacred palm nuts called Ikin or a divining chain known as an Opele Chain, strung with eight seeds from the Opele tree. These tools generate binary patterns that serve as spiritual codes, offering answers and guidance to life's challenges. The Ikin symbolize divine wisdom, and - much like the 0s and 1s that form the foundation of modern computer code - their configurations create binary outcomes. When the Ikin are manipulated or the Opele Chain is cast, they produce a sequence of two columns, each with four possible marks, totaling eight values. These values function as binary digits (like 0s and 1s / opened and closed), which correspond to one of the sacred oracle answers/chapters or verses called Odu Ifa. Each Odu contains specific verses, poems, stories, and rituals drawn from the vast Ifa literary corpus, the foundation of Yoruba spiritual knowledge. In total there are 256 principal Odu (with thousands of verses), each encompassing wisdom that addresses every possible experience and situation - past, present, and future. Interestingly, this mirrors digital computing: a single binary digit, or bit, is either a 0 or 1; grouped into eight, they form a byte, which can represent 256 possible values.

Much like artificial intelligence, which analyzes large datasets, identifies patterns, and makes informed decisions, the ancient Ifa Oracle is an advanced mathematical yet spiritual system. It uses a binary approach to provide practical and divine insights and solutions, demonstrating the sophisticated nature of this ancient African spiritual technology.

Mathematics: The conventional narrative of the history of mathematics often revolves around a Eurocentric perspective, heavily focusing on ancient mathematical systems in Greece and Arabia. However, this perspective is challenged by the unlimited mathematically precise colossal stone structures that still exist along the Nile River as well as two extremely ancient, significant mathematical artifacts that break away from Eurocentrism – the Lebombo bone, discovered in Southern Africa, and the Ishango bone, found at the origins of the Nile River on the border between Uganda and Zaire. Both of these carved tools are considered the world's two oldest mathematical objects.

The Lebombo bone, dating back at least 35,000 years, aligns with the historical evidence of iron ore mining in the region dating back 43,000 years. This small bone tool has been found with 29 notches carved into it and is considered one of Earth's oldest mathematical artifacts.

The Ishango bone, dating back around 25,000 years, is one of the oldest known tools for recording numbers and mathematical concepts. It features notches carved in groups, which were originally thought to be simple tally marks but are now believed to represent a more advanced understanding of mathematics. Some of the notches align with a system of counting and may even include a basic understanding of multiplication. Recent studies have revealed that the bone also likely served as a way to track lunar phases. Additionally, it is speculated that the bone might have been used by women to monitor their menstrual cycles, suggesting that Black women could have been the earliest mathematicians.

Navigation: Centuries before their European counterparts, Africans embarked on voyages to the Americas and Asia (not to mention the multiple waves of traveling Africans that populated the planet from top to bottom thousands of years earlier). Africa's intricate network of waterways served as bustling trade routes, inspiring various ancient African societies to construct an array of boats. These vessels ranged from small reed-based crafts and sailboats to more sophisticated structures equipped with multiple cabins and cooking facilities. From the Ancient Egyptians and Nubians to the black seafaring people of the Indian and Pacific Oceans to the Mali and Songhai civilizations, Black people constructed many boats of various sizes and dimensions capable of carrying many tons.

In the book "They Came Before Columbus," author Ivan Van Sertima discusses the African presence and contributions made in South and Central America. For example, the famous Olmec megalithic faces show the distinct, undeniable African presence in the Americas and elsewhere. The Atlantic Ocean currents, stretching from West Africa to South

America, played a crucial role in facilitating these maritime connections. Genetic evidence from plants, combined with descriptions and art from South American societies of that era, suggests that several African groups navigated across the oceans to establish a presence in South and Central America.

IMPORTANT NOTE: Numerous high-tech contributions made by indigenous Black people, whose ancestors originated from Africa, extend beyond the ones listed in this book. These ancient contributions, spanning Europe, Asia, the Pacific, and beyond, encompass a wide range of advancements, from indoor plumbing innovations to agriculture breakthroughs. Collectively, these high-tech contributions have had a profound impact, bringing about significant changes in civilizations worldwide.

African American Contribution to Technology

There are numerous African contributions to technology that haven't been detailed in this book. I am merely scratching the surface to underscore the fact that Black people were pioneers in implementing large-scale, technologically advanced projects. Throughout history, across the diaspora and into the modern era, Africans globally have consistently made significant contributions in the fields of computer science and technology while overcoming substantial barriers and obstacles.

One of the earliest known African American computer scientists is Katherine Johnson, a mathematician and aeronautical engineer who played a pivotal role at NASA (National Aeronautics and Space Administration) in the early 1960s. Johnson's calculations were instrumental in the success of the first American human spaceflight, as well as aiding in the advancement of the space program through the 1960s and 70s.

Another noteworthy African American computer scientist is Mark Dean. Dean, along with a team at IBM (International Business Machines), is credited with contributing to the development of the first PC (personal computer). Dean later led the development of various other key computer technologies.

Despite enduring the challenges of the Trans-Atlantic slave trade, over 400 years of slavery, Jim Crow, and segregation, Black people, often denied credit for their inventions and innovations, consistently remained at the forefront of technological advancement. The book "Black Inventors/ Crafting Over 200 Years of Success" by Keith C. Holmes delves deep into the common and frequently overlooked legacy of black creators and inventors.

A LIST OF A FEW HISTORIC AFRICAN AMERICANS IN INNOVATION AND TECHNOLOGY:

Granville T. Woods (1856 - 1910): Granville Woods was a revolutionary inventor from Columbus, Ohio, who is credited as the creator of fifteen different appliances related to electric railroads. He is most famously known for his multiplex "induction telegraph" invention, which enabled voice communication to be sent through telegraph wires, saving countless lives in the process by preventing accidents on the railroad. This has earned him the moniker of "Black Edison" (The New Imhotep would have been better) with a total of nearly sixty patents generated throughout his career.

Frank Greene (1938-2009): Frank Greene, a pioneer in technology, was coined as one of the "first Black technologists" who created high-speed semiconductor computer-memory systems in the 1960s and founded two tech companies. His invention of the world's fastest microchip in the 1960s laid the foundation for today's technology (including AI technology). He also started a venture capital firm, NewVista Capital, that supported minority and female-led companies. Today, he is remembered as a respected technologist and investor and is among the Silicon Valley Engineering Council's Hall of Fame inductees.

Otis Boykin (1920-1982): Otis Boykin was a Black inventor who helped revolutionize and improve electronics for everyday use - he held no fewer than 26 patents. His inventions included a wire precision resistor used in many types of equipment, including radios, televisions, IBM computers, and even military missiles. He also created a control unit for pacemakers, which significantly improved their effectiveness and saved lives.

Katherine Johnson (1918-2020): Katherine Johnson is one of the first African American women to work as a mathematician in the tech industry. Katherine Johnson taught math before applying to the Langley Research Center and joining NASA in 1953. Johnson calculated the flight path for the first NASA mission to space. Her calculations were crucial to the success of early missions such as Project Mercury and Apollo 11. In 2015, President Obama recognized Johnson's groundbreaking contributions to technology by awarding her the Presidential Medal of Freedom.

Clarence "Skip" Ellis (1943 - 2014): Clarence "Skip" Ellis, a Chicago native, earned the first Ph.D. in computer science for an African American in 1969 from the University of Illinois at Urbana-Champaign. He worked on software, hardware, and the development of the ILLIAC IV supercomputer. Clarence Ellis had a lengthy career at major tech firms like IBM, Bell Telephone Laboratories, Xerox, and others. At the Palo Alto Research Center, he led a dynamic team that invented Officetalk, the first office system using icons and Ethernet to enable remote collaboration. Ellis pioneered operational transformation, a field examining functionality in collaborative systems, now found in modern computer applications like Google Docs.

Annie J Easley (1933 - 2011): Annie J Easley began as one of just four Black people employed by the National Advisory Committee for Aeronautics (NACA). Easley started as a "human computer" (doing computations for researchers, analyzing problems, and doing calculations by hand). Still, with the advancement of machines, she adapted to learning assembly language and FORTRAN to become an accomplished pro-grammer. She is best known for her work on the Centaur rocket – a groundbreaking project that used a "first-of-its-kind" rocket with a unique fuel system never before seen in rocketry. Easley's work endures today as part of NASA's legacy and the legacy of aeronautics.

Mark E. Dean (Born March 2, 1957): Mark E. Dean is an inventor and computer engineer. He developed the ISA bus (The Industry Standard Architecture (ISA) bus was a hardware interface that connected peripheral devices to a computer's motherboard) and led a design team to make a one-gigahertz computer processor chip. He holds three of nine PC patents for being the co-creator of the IBM personal computer released in 1981. In 1995, Dean was named the first-ever African American IBM Fellow.

In recent years, many African Americans and other Black people have continued to make significant contributions to the field of computer science and technology. Next is a list of current black tech trendsetters pushing the game forward.

CURRENT BLACK LEADERS IN TECH:

Stacy Brown-Philpot, CEO and Fortune 500 Board Director: Stacy Brown-Philpot serves as the Founder & Managing Partner at Cherryrock Capital, an early-stage venture firm dedicated to investing in Black and Latinx entrepreneurs. As the former CEO of TaskRabbit, a leading task management network, she steered the company from a rapid startup to a global business, ultimately overseeing its successful acquisition by the IKEA Group.

With over a decade of experience at Google and Google Ventures, Stacy brought strategic expertise to the game. Her roles included leading global operations for key Google flagship products and serving as Head of Online Sales and Operations for Google India. Beyond her tech experience, she has a solid background in finance from her tenure at Price Waterhouse Coopers and Goldman Sachs.

Stacy's influence extends to her role as a founding member of SoftBank's $100mm Opportunity Fund, established to invest in Black and Latinx entrepreneurs. Her presence on the Board of Directors for HP Inc., Nordstrom, Noom, StockX, Joy, Black Girls Code, and The Urban Institute reflects her commitment to fostering innovation and diversity across various industries.

Andre Young, aka Dr. Dre, Creator of "Beats by Dre": As a founding member of the legendary rap group NWA, Andre Young stands out as a trend-setting pioneer with numerous years of innovation in the music and entertainment industry. Seamlessly blending his musical expertise with cutting-edge technology, he crafted his signature headphones, BEATS BY DRE. These headphones have gained immense popularity, culminating in the largest Apple acquisition of all time—a substantial 2.6 billion dollars.

Jessica O. Matthews, Founder and CEO of Uncharted Power Inc.: Jessica O. Matthews, acknowledged by the Obama administration, Fortune, Forbes, Harvard, and other esteemed entities for her brilliant inventions and visionary leadership, established Uncharted Power with the mission to address a significant challenge: sustainable city infrastructure.

Her journey began with the creation of SOCCKET, a power-generating soccer ball. Building on this success, Jessica O. Matthews founded Uncharted Power, a company dedicated to delivering essential services to underserved cities and communities. With 12 patents and patents pending, she is not only an innovative entrepreneur but also actively contributes to clean energy, women's empowerment, and STEM education through her service on multiple organizational boards.

Christopher Young, Executive Vice President, Business Development at Microsoft: Christopher Young currently holds the position of Executive Vice President – Business Development, Strategy, and Ventures at Microsoft Corp. In this role, he is tasked with formulating global business development strategies to propel growth across the company. Concurrently, Mr. Young serves on the boards of Snap Inc. and American Express.

Previously, he served as the CEO of McAfee, LLC, a leading independent cybersecurity company. His extensive experience also includes roles such as Senior Vice President and General Manager at Intel Security Group, where he spearheaded the initiative to spin out McAfee. Before that, Mr. Young was the Senior Vice President of the Security and Government Group at Cisco Systems, Inc., a prominent technology networking company.

Mr. Young has contributed to national security initiatives as a former member of the President's National Security Telecommunications Advisory Committee. Additionally, he has served as a director of the Cyber Threat Alliance and as a former member of the Board of Trustees of Princeton University.

Delane Parnell, Founder and CEO of PlayVS: As the Founder and CEO of PlayVS, Delane Parnell leads a $400 million venture-backed startup dedicated to constructing the infrastructure for high school esports. Before launching PlayVS, Delane achieved the distinction of becoming the youngest black venture capitalist in the United States while working at IncWell Venture Capital. Following this, he played a pivotal role in the early stages of Rocket Fiber, raising $31 million and contributing to retail strategy alongside the CEO. During his tenure at Rocket Fiber, Delane established Rush Esports, an esports team subsequently acquired by Team Solomid.

Under Delane Parnell's guidance, PlayVS has forged partnerships with 23 state high school leagues and numerous colleges and universities and extended its reach to additional youth organizations and events.

Melissa Hanna, Co-Founder and CEO of Mahmee: With a JD/MBA background, Melissa Hanna has accumulated years of leadership experience and channeled her passion for women's healthcare into Mahmee. This organization is dedicated to providing comprehensive support for mothers and newborns, spanning from pregnancy through the infant's first birthday.

Under Melissa's leadership, Mahmee has developed a platform that seamlessly connects mothers with healthcare professionals, ensuring a personalized and supportive experience. The organization goes beyond conventional services by offering expertise and ongoing educational support for families' emotional and physical well-being. Utilizing its proprietary HIPAA-secure technology, Mahmee's team of certified and trained maternity support professionals delivers proactive education and guidance to new families worldwide, all at a fraction of the cost.

Ime Archibong, Head of New Product Experimentation at Facebook: At the forefront of innovation, Ime Archibong, Vice President of product management and head of Messenger product at Meta, is driving the introduction of new apps and experiences to enhance the Facebook platform. His recent launch includes the implementation of default end-to-end encryption in Messenger, alongside other features aimed at providing users with richer and more secure personal messaging and calling experiences. As one of the highest-ranking Black executives at Meta, Ime is a vocal advocate for increased diversity, not only within his company but across the entire industry.

Phaedra Ellis-Lamkins, Founder of Promise: Founder of Promise, Phaedra Ellis-Lamkins, is leveraging technology to address criminal justice reform and modernize government systems. With a focus on streamlining the bail system, Promise brings outdated state and federal systems into the 21st century. Phaedra's work aims to drive solutions and improve outcomes for individuals exiting the criminal justice system.

Charley Moore, Founder and CEO of Rocket Lawyer: Identifying a need in 2014, Charley Moore founded Rocket Lawyer to connect tech start-ups and small businesses with quality legal counsel. Since its inception, Rocket Lawyer has served over 30 million individuals and 20 million businesses. Moore is an outspoken advocate for diversity in Silicon Valley, emphasizing the importance of building a pipeline of talented individuals from diverse backgrounds.

Angela Benton, Founder and CEO of Streamlytics: Angela Benton, Investor, Founder, and CEO of Streamlytics, has created a groundbreaking opportunity for users to reclaim their data through Universal Data Interchange Format technology. Streamlytics focuses on democratizing data collection by empowering consumers to own the data they create, pioneering the concept of "community-driven data." Angela's work has earned her recognition among Goldman Sachs' 100 Most Intriguing Entrepreneurs and Fast Company's Most Influential Women in Technology.

Lanny Smoot, Disney Research Fellow at Disney Imagineering: With a career spanning 25 years at Disney, Lanny Smoot, a Disney Research Fellow in Imagineering, is a world-class engineer and holder of 106 U.S. patents. His groundbreaking work on the HoloTile floor—a device facilitating movement in various directions—has revolutionized engagement in virtual reality and gaming, solidifying him as a distinguished innovator at Disney. With over 100 patents, he stands as Disney's most prolific inventor and one of the most prolific Black inventors in American history. In 2024, he became the First Disney Imagineer inducted into the National Inventors Hall of Fame (only the second from the Walt Disney Company to be inducted -the first was Walt Disney), marking a significant milestone and highlighting his groundbreaking contributions to the field.

Anna Makanju, Vice President of Global Affairs at OpenAI: As the Vice President of Global Affairs at OpenAI, Anna Makanju plays a pivotal role in advocating for good-faith regulation in the AI industry. OpenAI, a leader in the sector, is known for innovations like ChatGPT. Anna actively travels to discuss OpenAI's technology and contributes to the development and implementation of new AI regulations, making her a key figure on the cutting edge of AI advancements.

Alondra Nelson, Researcher at the Institute for Advanced Study and Policy Advisor: As the Director of the White House Office of Science and Technology Policy (OSTP), Alondra Nelson is responsible for responding to rapid changes in AI technology. She oversaw the release of the blueprint for an AI Bill of Rights in October 2022, providing a framework for AI makers and policymakers to ensure the public good in the rapidly evolving AI landscape.

Neil deGrasse Tyson: Neil deGrasse Tyson, an American astrophysicist, author, and science communicator, pursued his education at Harvard University, the University of Texas at Austin, and Columbia University. He served as a postdoctoral research associate at Princeton University from 1991 to 1994, focusing on star formation, exploding stars, dwarf galaxies, and the Milky Way's structure. Dr. Tyson is renowned for his significant contributions to popularizing astrophysical concepts and discoveries. In 2001, he was appointed by President Bush to serve on a twelve-member commission tasked with studying the Future of the U.S. Aerospace Industry, further solidifying his impact in the field of science and exploration. Tyson has since become a social media phenomenon, arguably making him the most popular astrophysicist of all time.

Robert F. Smith: Robert Frederick Smith stands as a distinguished American billionaire businessman and philanthropist, recognized as the founder, chairman, and CEO of Vista Equity Partners, a private equity firm exclusively dedicated to investments in software, data, and technology-enabled organizations. Revered as a trailblazer in sector-focused private equity investment, Smith has spearheaded over 610 completed transactions, representing a staggering $302 billion in aggregate transactional value since Vista's inception. Trained as an engineer at Cornell University, he earned his B.S. in Chemical Engineering in 1985. His exemplary leadership and example led to Cornell renaming the school he graduated from, the Robert Frederick Smith School of Chemical and Biomolecular Engineering, in his honor in 2016.

Smith's outstanding business accomplishments and philanthropic endeavors have garnered widespread recognition, including Harvard's prestigious W.E.B. Du Bois Award and the International Medical Corps Humanitarian of the Year Award. He also serves as the Chairman of Carnegie Hall and a board member of the Robert F. Kennedy Human Rights organization, contributing significantly to global humanitarian efforts. Additionally, he plays crucial roles as a member of the Board of Overseers of Columbia Business School and as an at-large board member of the Business Roundtable, further solidifying his influence and leadership in various areas.

David L. Steward: David L. Steward is a distinguished American billionaire businessman renowned as the chairman and founder of World Wide Technology, a prominent African-American-owned enterprise recognized as one of the largest in America. In 1990, he established World Wide Technology, Inc., which specializes in an array of cutting-edge technologies, including cloud capabilities, data center and virtualization, security, mobility, and networking technologies, alongside voice, video, and collaboration solutions. World Wide Technology stands as an $11.2 billion IT provider catering to esteemed clients such as Citi, Verizon, and the federal government, solidifying its position as a leader in the industry.

19 Keys (Jibrial Muhammad): Jibrial Muhammad, widely recognized as 19-Keys, emerges as a Global Thought Leader, pioneering various domains, including Web 3, Business, Mindset, Holistic Wealth, Tech, Metaphysics, and Financial Literacy, commanding a vast global following. His acclaimed show, High-Level Conversations on Earn Your Leisure, has garnered over 25 million views, solidifying his influence and reach. Additionally, as a technology analyst, 19 Keys educates the masses on Web 3, artificial intelligence, and innovation through his membership platform, Block World Order (BWO). BWO delves into Web 3.0 while enhancing mental, spiritual, and financial well-being. Members have access to live coaching, mentorship, events, cutting-edge algorithms, and courses, and engage with a private community of high-level achievers.

CURRENT AFRICAN TECH DEVELOPMENTS:

The technological proficiency of the black community should not be underestimated. I have consistently emphasized that solving global challenges requires perspectives outside the cultural models responsible for current world problems and issues. Genuine solutions must emerge from indigenous wisdom or those closely connected to it. If we keep seeking answers from the same people destroying the planet, the results will be the same, or we will accelerate disaster!

I recently asked ChatGPT to tell me, based on history, **what group of people could save the world from AI-driven destruction**. You might be pleasantly surprised to know that the response was heavily centered around indigenous and African wisdom and moral systems. Here is an abbreviated version of the response:

Indigenous Peoples (Worldwide) - Indigenous cultures around the world - from the Maasai in East Africa to the Lakota in North America to the Ainu in Japan - have long traditions of deep respect for balance, nature, and community over individual or corporate greed.

1. **African Diasporan Thinkers and Innovators** - From Ifa, Kemetic science, Ubuntu, and Pan-African ethics, the African Diaspora has long produced philosophical systems that center communal wellbeing, ethical responsibility, and ancestral intelligence.

2. **Ethical Technologists and Scientists** - Ethical thinkers and scientists often warn about AI's dangers and push for regulation, transparency, and alignment with human values.

3. **Youth and Grassroots Movements** - Young people and grassroots activities are leading the charge against extractive systems, AI being one of them.

4. **Spiritual Elders and Visionaries** - Dalai Lama, African High Priests, Sufi Mystics, Amazonian Shamans, spiritual elders hold frameworks that transcend materialism. Their teachings could anchor AI in consciousness, ethics, and humility.

Even ChatGPT understands the importance of new technology development coming through the minds of people rooted in systems that promote balance in life.

Presently, across Africa, the Caribbean, and South America, a remarkable cadre of young Black scientists, innovators, and entrepreneurs are creating groundbreaking and revolutionary technology. It must also be noted that other non-black indigenous people are also contributing amazing ideas and solutions to heal the world. Their ideas, if implemented, have the potential to revolutionize entire industries and help restore the health of the planet. Additionally, well-established entrepreneurs are making significant strides in the technology sector, contributing to its vibrancy.

Despite these advancements, there is a concern that others, particularly Western big business and other interests, will exploit and corrupt these innovations for profit. This tradition of exploitation involves stealing ideas, creating their own patents, and even sidelining the original creators. Such actions could lead to the same world problems and destructive patterns we aim to overcome.

Nevertheless, let's take a moment to recognize a few outstanding Black individuals who are making remarkable contributions to the tech industry.

Strive Masiyiwa: Zimbabwean billionaire businessman and philanthropist Strive Masiyiwa is the Founder of Econet Group, encompassing Econet Wireless and Cassava Technologies. Best known for his tenacity in establishing Econet Wireless, the leading mobile and fixed wireless platform in Zimbabwe, Masiyiwa has received accolades such as being named among Bloomberg's 50 Most Influential People, New African Magazine's 100 Most Influential Africans, and Mail & Guardian's 100 Africans of the Year in 2020. A board member of the Rockefeller Foundation, he co-founded the Alliance for a Green Revolution in Africa and, with his wife, Tsitsi, founded the Higher Life Foundation, dedicated to supporting vulnerable and orphaned children through education.

Charity Wanjiku: Forbes and the World Economic Forum recognize Charity Wanjiku, Co-Founder of Strauss Energy Ltd, as one of the leading women in global tech. Her solar company, Strauss Energy, pioneers green solutions to power rural communities in Kenya with patented solar tiles. These tiles, introduced before Tesla's similar product in 2017, provide unique solar systems equipped with a special meter that feeds unused electricity back to the national grid, generating income for households. Charity advocates for breaking STEM barriers for women and girls and contributing to architecture, entrepreneurship, and technology.

Mark Essien: Nigerian-born serial tech entrepreneur Mark Essien holds a Bachelor of Engineering (B.Eng) degree in Computer Hardware Engineering and an MSc degree in Computer Science. Founder of Hotels.ng in 2013, Essien aimed to make it the largest provider of travel information and reservations in Africa. Recognized by Forbes as one of the 30 Most Promising Young Entrepreneurs in Africa in 2015, Essien also founded the HNG internships in 2016 to train and recruit talented software developers across Africa.

Rapelang Rabana: Forbes' 30 Under 30 Africa's Best Young Entrepreneurs features Rapelang Rabana, Founder of Rekindle Learning. Her award-winning learning & development company provides mobile and computer learning applications for students and adults. Rabana's earlier venture, Yiego, was an innovative telecommunications firm developing some of the world's earliest mobile VoIP applications. A Global Shaper of the World Economic Forum, Rabana emphasizes education's power to create opportunities.

Tope Awotona: Tope Awotona, a disruptive Nigerian entrepreneur, is the founder and CEO of Calendly, a multi-billion-dollar scheduling platform. Calendly, born from Awotona's vision of simplifying scheduling, has become a prominent tool for high-performing teams and individuals. Awarded the Atlanta Business Chronicle's Most Admired CEO award in 2021, Awotona's journey involved risking his life savings and maxing out credit cards to fund Calendly in 2013.

Mary Mwangi: Mary Mwangi, a pioneer in Africa's fintech logistics space, is the Founder and CEO of Data Integrated, an ICT company based in Kenya. Specializing in financial solutions for African SMEs, Data Integrated focuses on Kenya's public transport system, leveraging tech to address industry challenges. Mwangi's MobiTillEpesi Smart Public Transport app revolutionizes fleet management, winning recognition such as the MEST Africa Challenge in 2018.

Sara Menker: Founder & CEO of Gro Intelligence, Sara Menker, employs artificial intelligence to forecast agricultural trends, highlighting the connections between the earth's ecology and the human economy. Menker, named a Global Young Leader by the World Economic Forum, is a fellow of the Aspen Institute and a Trustee of the Mandela Institute for Development Studies (MINDS). Her innovative approach stems from her upbringing in Ethiopia, inspiring a commitment to life and problem-solving.

Jay Alabraba: Jay Alabraba, Co-Founder and Director of Business Development at Paga, Nigeria's largest mobile payments company, has been a key figure in Africa's tech revolution. Alabraba, a pioneer in Nigeria's early fintech landscape, began his career at Microsoft before accepting the Charles P. Bonini Partnership for Diversity Fellowship from Stanford Graduate School of Business in 2004.

Father Godfrey Nzamujo: Father Godfrey Nzamujo is a scientist, founder, and director of the Songhai Regional Centre, established in 1985 in Porto-Novo, Republic of Benin, West Africa. Before founding Songhai, he served as a Research Fellow and Professor at the University of California, Irvine, Associate Pastor at St. Nicholas Catholic Church in Laguna Hills, California, and Associate Professor of Engineering at Loyola Marymount University in Los Angeles, California.

Founded on one hectare of land in a Porto-Novo suburb, the Songhai Centre draws its name from the historic 14th-century West African Empire. It serves as a training hub for agricultural entrepreneurship, emphasizing sustainable farming practices. As a self-sustaining, integrated farming institution, the Songhai Centre innovatively transforms waste into valuable resources. Its commitment to excellence and the Zero Emissions Research Initiative (ZERI) enables processes such as producing biogas for electricity and cooking, using maggots as fish feed, and turning fish waste into organic fertilizer for crops.

The Songhai Centre also produces a wide range of goods- soap, baked foods, juices, jams, syrups, animal feed, and more- that are sold in Songhai

Stores and distributed to wholesalers and retailers across Africa. The center's approach has expanded beyond Benin, contributing significantly to the future of sustainable agriculture in Africa and uplifting Black communities globally.

UP Next

In addition to the well-known African tech innovators mentioned above (please research for more details), numerous groundbreaking initiatives are quietly unfolding within local communities. Some of these emerging creators and their innovative concepts are gaining widespread attention on the internet, thanks to the ingenious nature of their eco-friendly solutions with substantial potential impact.

Cbokoudo: The Cbokoudo is a uniquely crafted two-wheeled vehicle from the Democratic Republic of Congo that resembles a bicycle. It is meticulously handmade using eucalyptus wood and serves various purposes, including transporting cargo, gasoline, coal, wood, and passengers.

Bamboo Bicycles of Ghana: The Ghana Bamboo Bike Initiative is creating bicycles out of bamboo. The creation of this technology is very lean on energy use. Bike production actually helps save the planet as compared to normal metal bicycle manufacturing. The company uses pre-used bamboo from construction sites across the country to make their bicycles, and even exports them around the world.

Maxwell Chicumbutso: Maxwell Chicumbutso, A young African scientist hailing from Zimbabwe, left school at the age of 14. Relying on blueprints and visions he attributes to divine inspiration, Chicumbutso faced various challenges securing patents for his ideas due to their unconventional nature, which appeared to challenged traditional physics laws. Among his groundbreaking creations, Chicumbutso engineered an electric car and a television that harnesses radio frequencies for energy, eliminating the need for batteries or recharging. Notably, he designed a fuel-free helicopter, and his drones are currently actively deployed in South Africa. Chicumbutso also created a household transformer that has the ability to amplify power a hundredfold.

John Wangare: John Wangare of Kenya created Magiro Mini Hydroelectricity, which generates electric power for 250 homes in his rural community. Wangare built the small power plant from leftover scrap materials anchored on a local water source, generating 250 kilowatts of electricity.

Shoe Charging Stations: The "charging shoe" stands out as an amazing African invention pioneered by Kenyan innovator Anthony Mutua. This groundbreaking technology utilizes power generated by pedestrians for mobile phone charging. The invention incorporates thin crystal chips seamlessly integrated into the shoe's sole. As an individual walks, the pressure exerted on the sole during each step generates electricity. The crystal chips then transfer this current through an extension cable, connecting the shoe to the phone in the pocket. Remarkably, the shoe continues to charge phones even after walking, releasing the stored energy when stationary. This innovative system allows individuals to walk or run without the need for continuous movement, ensuring energy generation and storage for later transfer to mobile phones.

Black To The Point:

Black people have always been on the cutting edge of technology and innovation since ancient times. Even without access and facing a variety of societal roadblocks, Black people are still able to achieve amazing results in the technology sector. It is time for us to collectively claim our space in the current technological revolution.

CHAPTER 5

MAAT vs Technology

Herodotus called the **Egyptians** "religious to excess, far beyond any other race of men."

It's not surprising that the Greeks revered Imhotep. Various European societies have a lengthy history of prioritizing mechanical technology over spirituality and civility within society (not fully understanding that Imhotep's achievements were fully rooted in his spiritual practice). As some Europeans progressed technologically, drawing from fundamental African contributions and others (Egypt to Greece and Greece to Roman, etc), and began to explore the world, they encountered various cultures less technologically advanced than their own. Unfortunately, this often led them to label these cultures as barbaric, savage, backward, pagan, and primitive, providing a pretext for attempts to wipe out or enslave indigenous peoples and appropriate their lands and resources worldwide.

The Greeks, thousands of years after Imhotep's death, held him in high regard. To them, Imhotep symbolized a "Divine Being", a prophet or God who introduced the healing arts, technical marvels, and advancements to the world. In reality, Ancient Egypt produced numerous geniuses besides Imhotep (Thutmose III, Queen Hatshepsut, Hesy-Ra, Ptah Hetep, Amen Hetep, Ahmose Nefertari, and countless other scribes, healers, priests, warriors, engineers, architects, and so on). For many generations, the Kamitic (Ancient Egyptian) society focused on nurturing the higher aspects of humanity's spirit, thus awakening the "divine genius" within.

The Greeks, under the tutelage of ancient Egyptians, utilized that knowledge to advance both technologically and to some degree, spiritually. In ancient times, predating the current modern era of "white supremacy", the trans-Atlantic slave trade, Arab conquest of Africa, and the Arabian African slave trade, it was widely acknowledged in the region that the "burnt-skinned" and "wooly-haired" Egyptians (as often described by Greeks, Romans, and others), Kushites, and Ethiopians were a most ancient educated people and experts in technology and spirituality.

For those seeking "higher learning," the destination was Kamit; there was no confusion about the contributions of Kamit, the Black Land full of Black people. However, purposeful confusion arose later as "non-Black", non-African external forces attempted to discredit Black African Egyptians, appropriating credit for achievements, religious concepts, and spiritual ideas they had not originated—a narrative that persists to this day.

While ancient Black African people were the foremost inventors of the advanced mechanical technology of their day (And everyone knew it), Africans have always held higher regard for their spiritual values, civility, harmony, and peace. While it's very cool to build a tall, mechanically complicated structure, it's even cooler to have positive, healthy, wealthy, sharing, harmonious, joyful people living their best, most peaceful, and divine lives. To this day, there still remain many happy, "traditional" African and non-African villages and communities of people worldwide who have decided to maintain their "low-tech" indigenous, technically minimalistic lifestyle despite offers from the outside technologically advanced world (the outside world is often making it hard for these indigenous people to exist). The fact is that the advanced spiritual lifestyle inspired and fostered the insight that led to the creation of the many mysterious technical marvels of ancient Egypt (The Pyramids, etc). Once the mathematical/technological foundations were laid, others simply copied, pasted, and applied the same or similar techniques to building structures in their societies. Most have been duplicating this process ever since, with very little "Spiritual cultivation" (nor giving credit to the Black originators) while looking down on others who choose not to incorporate mass technological mechanics. In indigenous Africa, on the other hand, there existed both highly technically advanced civilizations and societies with very little mechanical technology. Neither is better than the other...only different. The pyramids of ancient Egypt stand as a technical testament to the pursuit of spiritual enlightenment, while the Colosseum, arguably Rome's (Or even Europe's) greatest ancient technical landmark, stands as a testament to the moral values and spiritual aspirations of that society. Get the picture?

If society aims to foster positive, uplifting values among its people, owning an iPhone is not essential to achieving that goal.

Some societies placed a primary emphasis on the spiritual and harmonious aspects of life rather than solely focusing on mechanical technology. The ancient Egyptians (who's ancestors and wisdom were a culmination of even older African cultures), who laid the foundation for numerous cultures and subsequent world religions and civilizations, achieved technological mastery by first embracing and incorporating the spiritual values, principles and laws of Maat: Divine Law, universal truth, righteousness, justice, balance, and love (as outlined in "Maat: The 11 Laws of God" by Ra Un Nefer Amen). It was their mastery of the living the principles of Maat that allowed them access and provided the insight and mathematical brilliance necessary for the inception and execution of advanced megalithic and technological development.

Contrary to the notion of building technology first, the ancient Egyptians prioritized cultivating Maat. Their Kamitic cosmological system centered around Maat, emphasizing the interrelationship of the laws that govern all aspects of man's spirit and life. This perspective led them to recognize that living in accordance with the laws of Maat (God, the spirit, Nature, The Universe, balance, truth) took precedence over simply mastering left-brained mechanical technology, which is governed by "Sebek" (the intellectual faculty of the spirit. - See Metu Neter Volume 1 and Maat: 11 Laws of God by Ra Un Nefer Amen). Universal love, divine balance, order, and adherence to divine law and truth were the foundation for the society (No matter what Hollywood tries to tell you!).

Rather than using technology as an end in itself, the people of Kamit employed it as a tool to perpetuate the principles of divine law, love, and balance. This was evident in the spiritual messages etched in their highly advanced stone masonry, writings, temples, obelisks, statues, and spiritual lifestyle, reflecting a commitment to pushing forward such high ideas through technological engineering achievements. When walking down an ancient corridor in ancient Egypt, almost every image reflected a divine idea, as opposed to an advertisement for pizza, McDonald's, or cigarettes in today's society...Get it.

Sebek

It's essential to recognize that, despite the awe-inspiring nature of the pyramids and the Sphinx in ancient Egypt, these structures were ultimately erected as a testament to the spiritual way of life embraced by the people. The primary focus was not on showcasing the grandeur of the pyramids alone, but rather on highlighting man's / woman's potential to achieve such splendor by transcending his lower nature and revealing his divine, higher true nature, through adherence to the laws of Maat. Such a man or woman was considered an Ausar (Osiris by the Greeks). The ultimate aim of every individual was to become an Ausar - one who embodied the oneness of god and thereby gained access to divine power. The ancient Egyptians taught that the faculties of the human spirit, which govern mechanical skill, intellectual reasoning, and even what we now call Artificial Intelligence, were attributed to the spiritual faculty known as Sebek. Yet Sebek was always placed lower in the spiritual hierarchy than the divine principles of Maat - truth, divine law, and order. While intellect, analytical thought, and technological advancement were acknowledged as vital to survival and often extraordinary in achievements, they were never meant to overshadow higher spiritual development. The true purpose of the intellect was to serve as a tool in helping one live in alignment with the laws of Maat, also called the law of god (Neb Er Tcher)

In contrast, in much of modern society, intellect and technology are exalted for their own sake, with little regard to divine law. What results is not balanced but what could be described as "spiritual lawlessness" as well as "technological lawlessness". In ancient Africa, the purpose of life was the attainment of unity with god. But a civilization that elevates technology and segregative thought above the divine inevitably breeds disorder, corruption and chaos. Thus, while the Sebek faculty is indispensable to human survival, it is also the very source from which evil and chaos may arise when it operates outside the guidance of The Laws of Maat.

Many other cultures achieved technological advancements primarily by imitating, replicating, and building upon what was already conceived and created by the people of Africa and Kamit (e.g., Greece, Rome, Persia, etc.). With Maat as its foundation, ancient Africans proceeded to establish a society that embodied her laws in every aspect of life, including mechanical technology.

It must also be noted that it is possible to attain a certain level of technological advancement through scientific manipulation without being deeply rooted in the practice of living the laws of Maat. The crucial question is, without Maat (or a similar system of living that allows people to examine and balance interconnected relationships for good), what kinds of technology will be developed, and what will be the ultimate impact on the well-being of society, the planet, and the connection to nature and the universe?

Africans grasped the idea early on that great technology, especially when utilized inappropriately, does not result in "progress" or contribute ultimately to personal improvement. The development of more powerful bombs in smaller packaging does not guide a society toward enlightenment. While advanced technology may enhance the efficiency of certain tasks (some giving rise to more intricate problems and issues like global warming), genuine progress and societal progress stem from the positive development and advancement of the hearts, minds, and spirits of its people. A healthy society would not trade the long-term health and well-being of its people for short-term material "progress". The process of personal spiritual development extends beyond individuals, contributing to the upliftment of communities, nations, and the planet.

I don't wish to suggest that ancient Africa or ancient Egypt were utopias - no culture is without its challenges. The notion of a utopia is a fantasy that can lead to unrealistic perceptions of ancient Africa or anywhere else. Rather, my point is that by embracing the Maat tradition, which emphasizes law, love, and order as guiding principles, these societies established a foundation that encouraged people to strive for truth, righteousness, and oneness. This focus inspired many individuals to strive for a holistic view of life by cultivating their collective divinity, thus creating solutions that benefited themselves, their communities, and the world around them.

The Tech Problem

The challenge with technology lies in the misconception that the creation of new gadgets equates to mankind's advancement. This misunderstanding arises because we often measure our "progress" through a Western lens. The assumption that possessing a superior toaster, car, spaceship, gun, or phone means true "progress" for "human beings" is far from the truth.

In a materialistic society, where the paramount value is placed on the rapid marketability and sales of products, it is logical to misguidedly associate "true progress" with various mechanical innovations. The current popular world view fosters an environment where might is right, as seen in advances in military technology, and the one with the biggest arsenal, whether guns, bombs, weapons, or an army of sophisticated attack AI robots, prevails. Meanwhile, as the products advance and become available at cheaper costs, the people whose resources are being used to create the new technology are hardly ever truly compensated, and many are totally left out of the so-called "progress" paradigm.

However, it doesn't require much intelligence to recognize the detrimental impact of fossil fuel waste production on our environment in just

50 years (while Kamit endured for over 3,000 years). While the automobile, fueled by a competitive market and fossil fuels, seemed like a fantastic idea on the surface, the early developers overlooked, lacked wisdom and foresight, were driven by greed, didn't know, or simply didn't care about the ultimate environmental cost. The early inventors seemed to have lacked a sense of Maat, which would have helped them to analyze in advance and weigh the pros and cons of the environmental effects.

A recent viewing of the film "Oppenheimer" depicted a group of inventors exhilarated about harnessing the power of the atom to create the atomic bomb (leading to the nuclear bomb). Yet, in one scene, the film hinted that this creation might open the door to the eventual destruction of all life on Earth. Despite such concerns, the most celebrated scientists in America participated in the invention of a weapon that obliterated Hiroshima and Nagasaki, killing hundreds of thousands instantly (the official reasoning was that it would end World War II and save lives). This use of new technology, without thoroughly assessing (and or weighing) the options and consequences, has irrevocably altered warfare and could, at any moment, lead to WORLD WAR III (if even accidentally) and our collective demise. Once the "electric cat" is out of the bag, it's hard to turn back.

Enter Maat, the ancient African concept embodying the interconnectedness of all things, both seen and unseen. This philosophy posits that all relationships are governed by laws rooted in truth and love, maintaining an ultimate divine oneness, order and balance (even if you are ignorant of that relationship). The ancient Black African scientists and priests consistently pondered, "What impact will this creation or decision have on the world? How will it influence the overall balance of life?" If the answer proved negative, the idea was either abandoned or modified long before any widespread creation, production, or implementation.

In contrast, the historical Western approach seems to be, "If I can invent it and sell it, then I will create it! Let's get rich! We'll deal with the consequences (or not) after selling a few billion of these bad boys!" Even if, for instance, it entails establishing harsh production environments where children work in squalor and are exposed to toxic chemicals daily. Even if it involves mass slavery. Even if it leads to air pollution affecting our grandchildren. Even if it results in creating a bomb capable of extinguishing all life on Earth in 30 minutes. Let's get rich! How many James Bond films have we seen where extremely "intelligent", "civilized" and "sophisticated" geniuses are on the verge of destroying the world through the use of some extremely advanced, technically complicated apparatus just to make a buck?

Regrettably, once the symptoms of decay manifest, the damage is often irreversible. Therefore, the ancient Black scientist consistently erred on the side of Maat from the inception of any project, concept, or creation. The idea was to undertake endeavors that maintain spiritual laws and balance rather than disrupt them.

The Microchip

With regard to the creation of computing technology, it must be stated that AI tech would not exist as we know it if it were not for the creation of microchip technology. A microchip, often smaller than a fingernail, houses an integrated circuit containing computer circuitry—a groundbreaking invention regarded as one of mankind's most significant technological advancements. Nearly all modern technology products utilize chip technology. In 1959, scientists Jack Kilby of Texas Instruments and Robert Noyce of Fairchild Semiconductor Corporation secured patents for pioneering this innovation.

A microchip comprises interconnected electronic components, such as transistors, resistors, and capacitors, etched onto a minuscule chip of semiconducting material, like silicon or germanium. Microchips serve various functions, from powering the logic component of computers, known as microprocessors, to storing computer memory or RAM chips.

Advancements in chip technology and size have played a pivotal role in propelling computing forward rapidly. Microchips are now ubiquitous in various applications, including AI-driven devices and smart-phones, facilitating internet access, video conferencing, gaming, GPS tracking (did you know an African American woman mathematician by the name of Gladys West was instrumental in laying the foundation for GPS?), virtual reality, data transfer, and the global economy. Moreover, they are essential in healthcare for expediting the diagnosis of diseases like cancer and potentially aiding in curing conditions like dementia and Parkinson's disease.

Microchips have come a long way since 1959. In 1999, Mark E. Dean, an African American personal computer pioneer, made a breakthrough in data processing while working for IBM. Mark and a team of engineers created the very first gigahertz chip. Prior to then, computer processors were operating in kilohertz (one thousand cycles per second) and mega-hertz (one million cycles per second). With Dean's game-changing piece of technology, computers were able to perform one billion calculations per second. This revolution changed computing altogether and laid the framework for today's technology.

Recently, Jensen Huang, CEO of Nvidia and a leader in artificial intelligence computing, announced the development of newer, larger next-generation AI chips—the Blackwell Chip. Nvidia's Blackwell chip, the Blackwell Ultra, and others are graphics processing units (GPUs) with microarchitecture that are designed to train large AI models. The Blackwell chips are part of Nvidia's Blackwell platform, which delivers significantly faster information processing, poised to transform AI and enhance efficiency exponentially. While introducing the chips at a recent event, Jensen displayed Humanoid Robots and discussed how the Black-well chip and other Nvidia products and systems are ushering in a new and profound AI-driven industrial revolution. Nvidia, however, is not the only player in the semiconductor space. Taiwan Semiconductor (TSMC), Samsung Electronics, Intel Corp, Broadcom Inc., SK Hynix Inc., Applied Materials Inc., Semiconductor Manufacturing International Corporation, Hangzhou Silan Microelectronics, and others, all are pushing the chip game forward, attempting to keep a competitive edge in the artificial intelligence landscape.

While understanding the intricacies of chip technology may not be necessary for everyone, it is important to at least recognize that their widespread usage and creation are crucial to the survival of the tech and AI industries. The growing demand for computing components suggests an increase in the need for raw materials, resources, and elements used in chip production. This poses many challenges, particularly in Africa, where a large percentage of the necessary elements are sourced. It is the criminal and corrupt resource extraction of the raw materials that have laid waste to the practice of the Laws of Maat, and are destroying millions of Black lives in the name of profits for multi-national tech giants.

The Cobalt Mines of the Congo:

More than 70% of the world's cobalt, used in batteries for electric cars, computers, cell phones, and various technological devices, is extracted in the Democratic Republic of the Congo (DRC). The DRC is acknowledged as one of the most, if not the most, mineral-rich countries globally. With the rising demand for cobalt, particularly due to the surge in electric car sales, especially in Europe and America, incentivized by generous environmental bonuses, the World Economic Forum's Global Battery Alliance forecasts a fourfold increase in cobalt demand for batteries by 2030.

A significant issue arises as Western and Asian companies profit substantially from mining activities while impoverished Congolese miners, primarily Black individuals (numbering around 255,000 and increasing), including many children (around 40,000), endure lives of poverty, squalor, and death. They continue to mine under toxic, unregulated, extremely harsh, life-threatening, and environmentally destructive conditions. These mines pose a literal threat to the lives of the Black miners, so much so that the minerals extracted are even labeled as "conflict minerals." Because of this destabilized environment, the DRC has suffered a loss of over 5 million lives in the deadliest conflict in modern African history since 1998. Western nations have contributed to the instability of the Congo, primarily because of their reliance on cheap raw minerals. Through the often irresponsible extraction of gold, cobalt, coltan, copper, uranium, diamonds, and others, western nations continue to advance themselves technologically and make money through the sale of hi-tech products.

Lithium Mines:

Africa is poised to boost its lithium production in the coming years, with a significant portion originating from Zimbabwe. As the lightest known metal, lithium finds widespread use in electronic devices, ranging from mobile phones and laptops to cars and aircraft. However, lithium extraction poses dangers, leading to soil degradation, water shortages, biodiversity loss, damage to ecosystem functions, and an escalation in global warming. Zimbabwe also loses over a billion dollars in mineral proceeds to foreign companies, perpetuating the exploitation of its resources. In addition to Black people working as employees rather than landowners, they endure unhealthy conditions and receive meager wages while foreign investors and entities reap extremely high profits.

Gold:

Gold mining in African countries is linked to human rights abuses, extensive pollution, vast open pits, detrimental community health impacts, worker hazards, and, in many instances, environmental harm. Research indicates that gold mining acts as a 'resource curse' for countries like Guinea, Niger, Zambia, and Togo, becoming a dual curse for local communities by causing both physical displacement and the loss of traditional livelihoods. Apart from its use in jewelry, gold is a crucial resource for producing cell phones, computers, and various other technologies.

Iridium:

Haiti boasts one of the world's largest iridium reserves, standing as the second-largest holder after South Africa. Despite its rarity and challenging nature, iridium is abundant in Haiti's southeast region. This metal, currently more valuable than gold and Bitcoin, is indispensable due to its resistance to high temperatures and corrosion, particularly in manufacturing aircraft engines and deep-water pipelines. In a country grappling with a state of emergency, the question arises: How would iridium mining impact the lives of Haiti's Black population? Could it lead to overnight wealth for the nation, or would it exacerbate the struggles faced by those seeking riches at the expense of Black communities worldwide?

Examining just a few elements essential for creating computing components reveals the negative impact their production has on Black communities globally. African and other Black countries not only possess abundant minerals crucial for creating technology but also have a long history of being exploited by foreign interests, internal puppet governments, and other companies and entities interested in resource extraction that detrimentally affect the safety, health, and economic prosperity of Black people.

Is the convenience of owning the latest high-tech gadgets worth the harm caused to Black lives worldwide (ABSOLUTELY NOT!)? How can the system be balanced (MAAT) to ensure fair wages, national wealth, environmental protection, healthy working conditions, and prosperity for Black people while we still enjoy our favorite gadgets and devices like flat screen TVs in every room (Or can it?)? It's time for an end to the wars, violence, and divide-and-conquer tactics that we have been falling victim to for so many years! It's time for Africa and Black people to benefit and prosper economically and environmentally from the sale of their talent, hard work, and natural resources!

The true involvement of Black people (politicians, tech leaders, entrepreneurs, sports and entertainment figures, etc) in technology creation can help drive a positive shift from the current situation into one that fosters pride, safety, healing, wealth generation, and prosperity.

Black To The Point:

As we embrace new technology, always be conscious of the fact that our new gadgets do not make us better as people, just perhaps more efficient or maybe even more distracted. Let us choose truth, justice, and love as we address the injustice in the creation of our algorithms and gadgets.

CHAPTER 6

Impact of AI On The Black Community

Having established the fact that Black people have always had a significant hand in the creation of advanced technology, let's now turn our attention to the impact current AI trends are having on the black community. AI's effects are both positive and negative. I have compiled an easy-to-review list of some of the most impactful:

- **Unequal access to education and training:** Black workers are less likely to have access to the education and training that would help them acquire the skills needed to work in jobs that are less likely to be automated. This can put Black people at a disadvantage when it comes to securing jobs in the AI-driven workforce.
- **Concentration in low-wage jobs:** Black workers who are employed in low-wage jobs are more vulnerable to automation and are therefore at risk of losing their jobs. For example, many Black workers are employed in industries such as retail and food service, which are increasingly being automated through the use of robots and other technologies. This can lead to increased economic inequality.
- **Bias in AI algorithms:** AI algorithms can perpetuate and amplify existing biases and discrimination. For example, if the data used to train AI systems reflects existing racial disparities, these systems may reinforce these disparities by making biased decisions that impact Black workers disproportionately.
- **Lack of representation in the tech industry:** Black people are underrepresented in the tech industry, which is responsible for developing and deploying AI systems. This means that Black workers are less likely to have a voice in the development of AI and in decisions about how AI will impact the workforce.

These factors, along with others, contribute to the increased risk faced by Black workers in the face of AI automation. To mitigate this risk, policymakers, educators, and tech leaders must work together to ensure that Black workers have the skills and support they need to succeed in the AI-driven workforce.

1. Bias in AI Systems: AI systems can perpetuate and amplify existing biases in society, particularly in areas such as criminal justice, where AI is used to make decisions about bail, sentencing, and parole. Black people are often unfairly targeted by these biased AI systems, leading to further racial injustices.

- **Bias in AI algorithms:** If the data used to train AI systems reflects existing racial disparities in the criminal justice system, these systems may reinforce these disparities by making biased decisions

that impact African American suspects and defendants disproportionately.

- **Predictive policing:** AI can be used to predict where crimes are likely to occur, but if the data used to train AI systems is biased, it may continue to lead to biased policing practices that target African American communities unfairly. Many home security and surveillance AI-driven tools can reinforce racism and lead to false arrests.

- **Sentencing algorithms:** AI algorithms can be used to make decisions about sentencing, such as recommending a sentence length or determining whether a defendant is a flight risk. If these algorithms are based on biased data, they may recommend sentences that are harsher for African American defendants than they are for defendants from other racial groups.

- **Facial recognition technology:** AI-powered facial recognition technology is increasingly being used in the criminal justice system, but it has been shown to have significant accuracy problems for people of color, particularly African Americans. This can lead to false arrests and wrongful convictions.

2. Healthcare Disparities: AI has the potential to improve healthcare outcomes for Black communities, but it can also exacerbate existing disparities if the data used to train AI systems is not diverse and representative.

- **Improved medical decision-making:** AI algorithms can analyze large amounts of medical data to provide healthcare providers with insights into patient health. This can improve the accuracy of diagnoses, reduce the time required for treatment, and ultimately lead to better patient outcomes.

- **Predictive analytics:** AI can help healthcare providers identify patients at risk of certain health conditions, such as heart disease or diabetes. This can allow providers to take preventive measures to improve patient health.
- **Personalized medicine:** AI can help healthcare providers create personalized treatment plans for patients based on their individual health histories and genetic information. This can lead to improved patient outcomes and reduced side effects from treatments.

3. Negative Healthcare Consequences:

- **Lack of data diversity:** AI algorithms require large amounts of data to train and function effectively. If the data used to train AI systems is not representative of the full diversity of the patient population, it can lead to biased decisions that harm Black patients.
- **Unequal access to technology:** Black communities are often disadvantaged when it comes to access to technology and digital infrastructure. This can limit our ability to access the benefits of AI in healthcare, such as telemedicine and remote monitoring.

4. Education: AI has the potential to improve educational outcomes for Black students, but it can also widen existing disparities if Black students do not have knowledge of and access to AI technology and resources.

- **Personalized learning:** AI can analyze students' strengths and weaknesses and provide tailored recommendations for educational materials and activities that meet their individual needs. This can help Black students develop a stronger foundation in subjects they struggle with and excel in areas they are interested in. AI tools can also address cultural differences that influence educational outcomes, helping Black students become better prepared and achieve greater success in academic settings.

- **Early intervention:** AI can analyze student performance data and predict academic risk, allowing educators to intervene early and support students before they fall behind. This can help Black students overcome obstacles that might otherwise hold them back and achieve academic success.

- **Virtual tutors:** AI-powered virtual tutors can provide students with immediate feedback, explanations, and additional practice opportunities. This can help Black students reinforce their understanding of key concepts and avoid confusion that can lead to frustration and disengagement.

- **Improved access to resources:** AI can help match students with resources and materials that are best suited to their individual needs, regardless of location or availability. This can help Black students overcome access barriers that might otherwise limit their opportunities and success.

- **Improved teacher support:** AI can help teachers better understand the needs and progress of their students, allowing them to provide more targeted and effective support. This can help Black students receive the support they need to succeed and reach their full potential.

A few ways AI will affect the Black community in the future include:

1. **Job Creation:** AI has the potential to generate new employment opportunities, presenting economic advantages for Black communities. I've recently observed Black Instagram users selling pictures and posters featuring captivating art images generated in seconds by AI algorithms. While I empathize with the average Black visual artist, I'm also enthusiastic about the new entrepreneurs this technology will foster. Let's be honest, most of us opt for Uber and Lyft over traditional cabs. Uber disrupted the cab business, transforming it in many ways. Conversely, this disruption created an environment for innovative entrepreneurs to introduce new rideshare business concepts as well as have access to flexible hour, driving job opportunities. Unfortunately, it looks like Waymo, the autonomous, AI rideshare service, could disrupt the human-dri-

ven rideshare services very quickly. As AI takes over, people have unlimited opportunities to create new ways to disrupt the industry and provide new business models and cash-generating opportunities.

2. Improving Healthcare Outcomes: AI has the potential to improve healthcare outcomes for Black communities, particularly in areas such as chronic disease management and mental health. New AI-driven technology is diagnosing breast and other cancers long before they spread. These tools are also assisting in operations and cancer tumor removal in a more precise way, which leads to quicker healing. Who could better help develop the next AI-driven wave in health care for illnesses that disproportionately affect Black people? You guessed it. It is literally our responsibility, first and foremost, to address our health issues. Imagine more holistic, alternative approaches to healthcare that could also be expanded upon by proper AI programming!

3. Addressing Bias: As AI technology advances, it is possible that new tools and techniques will be developed to detect and prevent bias in AI systems, which could lead to more equitable outcomes for Black communities. These are the kinds of tools that only Black AI creators can ensure are properly coded and created.

It is important to note that the impact of AI on Black communities will depend on many factors, including access to technology, resources, and representation in the AI industry. By being informed and engaged with AI, Black communities can help shape its development and deployment in equitable ways that benefit Black communities. After examining the truth that many jobs Black people currently have will be lost to AI, it is of utmost importance for Black people to quickly examine the current trends and build for a bright future.

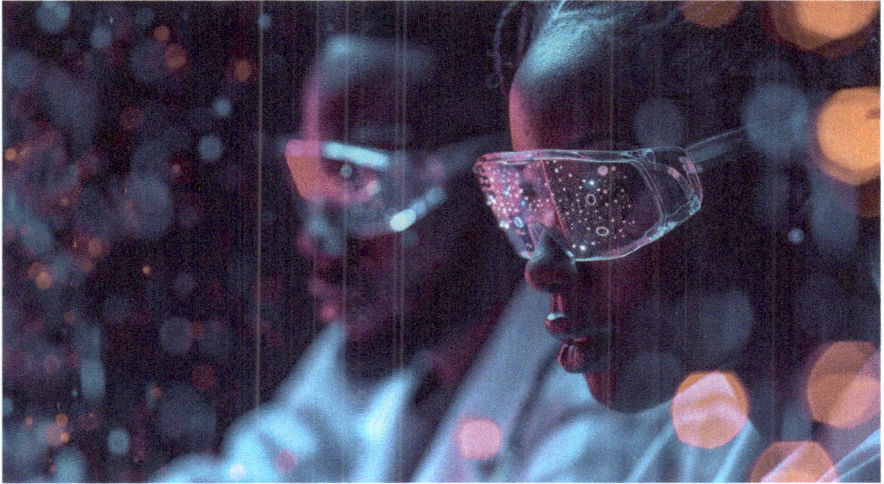

A Few Familiar Major Companies That Utilize AI Technology:

Several major companies have traditionally been the primary players in the development of AI technology for years. We all have used these brands in one way or another and have therefore contributed to their AI development, whether we knew it or not. Some of the most notable include:

Google: Google remains at the forefront of AI through its advancements in natural language processing, computer vision, and AI research. Its tools, like Google Cloud AI, provide scalable AI solutions for enterprises. Google Bard, their conversational AI model, competes directly with other AI chatbots like ChatGPT. Google also integrates AI into products like Google Assistant, AI mode, Search, and its Android ecosystem, and it has launched Gemini, a next-gen AI model combining the strengths of its DeepMind and Google Brain teams. Google recently announced AI enhancements with creative tools like Veo 3 (Googles latest AI video model that produces realistic high-resolution videos with audio and dialogue), Imagen 4 (updated image generation and integrated workflow app), Flow (an Ai powered suite that allows users to generate storyboards, film previews and even short films based on prompts), and many others. TensorFlow continues to be a major open-source platform for machine learning, used widely by developers and researchers. In autonomous driving, Waymo, a subsidiary of Alphabet, continues to expand its self-driving car technology.

Microsoft: A major player in AI with various products and services like natural language processing, computer vision, security and threat detection, chatbots, virtual assistants, and cloud-based machine learning tools. Microsoft has a strong partnership with OpenAI; Microsoft has integrated OpenAI's models (like GPT-4) into products such as Azure AI, OpenAI Service, Microsoft Copilot / Microsoft 365 Copilot (embedded in Office products like Word and Excel), GitHub Copilot, and Microsoft Teams. Azure AI offers comprehensive AI tools for businesses, including cognitive services and machine learning models. Microsoft is also making strides in AI for gaming through Xbox and AI-enhanced experiences on Windows 11.

Amazon: Amazon Web Services (AWS) continues to be a leading platform for cloud-based AI services, including SageMaker for machine learning and Rekognition for image and video analysis. Amazon also uses AI to create personalized recommendations based on customer shopping history. Amazon uses AI in its logistics and warehousing to optimize operations with over a million robots that handle tasks like sorting, picking and moving packages. Alexa remains a popular AI-powered virtual assistant in smart homes, and the company is focusing on increasing its generative AI capabilities for customer service and personalized shopping experiences.

Meta (Facebook): Meta is heavily investing in AI for the metaverse, virtual reality, and augmented reality. Its Meta Quest 3 headset integrates AI to enhance mixed-reality experiences. The company also invests in large-scale AI models for content moderation, recommendation algorithms, and language models, including open-sourcing models like LLaMA (Large Language Model Meta AI). Meta has implemented labeling for AI-generated content across platforms like Instagram and Facebook, and its AI-driven ad targeting continues to be a major revenue driver. Meta AI is an assistant integrated across platforms like Facebook, Instagram, and WhatsApp that you can interact with through text and voice. Meta AI has video, image generation, and other cool tools for business and personal use. Mark Zuckerberg, the CEO of Meta, has also recently announced the creation of multiple "titan clusters" of AI data centers, one of which is being described as being nearly the size of Manhattan.

IBM: IBM remains focused on enterprise AI, with its Watson platform evolving to include industry-specific solutions in healthcare, finance, and customer service. IBM's AI strategy emphasizes explainability and trustworthiness, aiming to make AI solutions transparent for businesses. IBM is also innovating in areas like AI quantum computing, which could significantly impact AI development.

Apple: Apple uses AI to enhance user experiences across its ecosystem, including Siri, on-device processing for privacy, and features like image recognition in the Photos app. **Apple Intelligence**: This new feature is available on iPhone, iOS 18, iPadOS 18, and macOS Sequoia. It uses AI to help users with tasks like writing, getting things done, and expressing themselves. Apple Intelligence can understand personal context across Apple devices to make recommendations and generate results. It's custom silicon, like the M1 and M2 chips, integrates advanced machine learning capabilities directly into devices, allowing for fast and efficient AI processing. Apple runs machine learning algorithms directly on devices like phones, watches, and speakers. This allows for faster processing of functions like Face ID logins, camera features, augmented reality, and battery life management. Apple has been investing in augmented reality (AR) and recently announced the Vision Pro, a mixed reality headset that integrates AI to enhance AR experiences.

Tesla: Tesla's Autopilot system uses AI to help cars drive themselves. The system uses cameras, radar, and ultrasonic sensors to monitor the car's surroundings and make decisions about how to respond. Tesla continues to lead in AI for autonomous driving with its Full Self-Driving (FSD) software, which aims to enable its vehicles to navigate complex environments using computer vision and deep learning. The company uses massive amounts of driving data to train its AI models, aiming to make self-driving a universal reality. Tesla's AI Day events provide insights into their AI advancements, including the development of XAI, Tesla Bot, Optimus, Cybercab, Robovan, and other products.

X (formerly Twitter): Under new leadership, X is incorporating AI for content recommendation and moderation, aiming to improve user experience and engagement. The platform is exploring AI-driven monetization features for creators and advertisers. X is also utilizing a chatbot called "Grok" that can interact with users using natural language processing. Recent focus areas include exploring generative AI capabilities for content creation and user interaction.

Uber: Uber leverages AI to optimize ride-matching, pricing, and route recommendations for drivers and riders. Its AI systems help predict demand patterns and allocate resources efficiently. The company is also advancing self-driving technology through partnerships and its autonomous vehicle research arm, aiming to reduce costs and improve safety.

IBM Cloud: IBM Cloud offers AI-infused cloud solutions, with an emphasis on hybrid cloud and AI integration. Its AI capabilities include automation, predictive analytics, and advanced data analysis, supporting various industries in their digital transformation. IBM's focus on AI ethics and data privacy aligns with its offerings to enterprises looking for secure AI solutions.

NVIDIA: NVIDIA is a leader in AI hardware, known for its powerful graphics processing units (GPUs) that are essential for training and running AI models. Its GPUs are widely used in data centers, autonomous vehicles, and AI research. NVIDIA's software platforms, such as CUDA and the NVIDIA AI Enterprise suite, help businesses and developers deploy AI applications efficiently. NVIDIA's AI tools, like Omniverse, are also being used for digital twin simulations and metaverse development.

OpenAI: Known for developing the GPT series, including GPT-4, 5 and beyond, OpenAI is at the forefront of generative AI, powering a wide range of applications from chatbots to content creation. Its models are integrated into various platforms, including Microsoft products like Azure and Office 365. OpenAI's focus on safety and alignment with human values in AI is a core part of its strategy, alongside developing the ChatGPT product line, which has been adopted by millions of users globally. ChatGPT has expanded beyond chat to incorporate other apps like DALL-E to create images from words, Python/Code Interpreter to solve math problems and analyze data, and Zapier to automate tasks between apps like Gmail, Google Sheets, and Slack. Please take the time to explore some of the other GPTs associated with ChatGPT (Canva, Wolfram, SciSpace, Video GPT by VEED, AI Humanizer, ChatPRD, Song Maker GPT, Writing Assistant, Presentation, TurboScribe, Whisper Transciber, Coloring Book Hero, etc.).

Salesforce: Salesforce has embedded AI into its Customer Relationship Management (CRM) platform with Einstein AI, which provides predictive analytics, natural language processing, and personalized recommendations. Its recent focus on generative AI capabilities, such as Einstein GPT, helps businesses automate customer service, sales, and marketing activities. Salesforce has integrated AI to help enterprises harness their data more effectively for strategic decision-making.

Baidu: As a leading AI company in China, Baidu is known for its investments in autonomous driving through Apollo, an open-source self-driving platform. Baidu also leverages AI in its search engine and offers cloud-based AI services through Baidu Cloud, similar to Google and AWS. Baidu's Ernie Bot is a competitor to ChatGPT in the Chinese market, and the company continues to push the boundaries of AI applications in language processing, computer vision, and smart city solutions.

Adobe: Adobe has integrated AI across its suite of creative software, such as Photoshop, Illustrator, and Premiere Pro, through Adobe Sensei, its AI and machine learning framework. Adobe is also exploring generative AI capabilities, allowing users to automate complex creative processes like image editing, video production, and content generation. The recent addition of Firefly, Adobe's generative AI platform, aims to assist creators in generating high-quality, custom images, videos, and 3D assets quickly.

Some of The Hottest
Easy-to-Use New AI
Technology:

In recent years, and particularly amid the COVID-19 pandemic, numerous AI apps have gained popularity, becoming both accessible and user-friendly. I'm sure you have encountered some of this technology without even realizing it. When I started writing this book, very few Generative AI apps were on the market. Nowadays... everybody is an AI expert.

Hot New Fast-Moving AI
Tools / Brands

(That may have disappeared or been replaced by the time you read this book. Lol)

ChatGPT (OpenAI): ChatGPT is part of OpenAI's GPT family, currently in versions like GPT-5. It's a powerful generative AI tool capable of understanding and generating human-like text. It can handle tasks such as answering questions, creative writing, translation, and more. It powers various applications, including AI-driven content creation and customer support.

SORA, SORA 2 (OpenAI): A fairly new AI tool by OpenAI, SORA allows users to generate moving images based on prompts. It's a significant advancement in generative AI, offering applications in video production, animation, and the film industry. SORA enables creators to turn written concepts into visual scenes, posing a shift in how storytelling and content creation might evolve. Since the advent of Sora, many AI video-generating apps have come to market.

Siri and Alexa: Both are AI-powered virtual assistants. Siri (Apple) and Alexa (Amazon) use natural language processing (NLP) and speech recognition to interpret user commands, control smart home devices, and provide answers to various questions. They continue to improve in understanding context and offering personalized responses.

Tesla Autopilot: Tesla's autonomous driving system uses AI for real-time image recognition, path planning, and decision-making to enable its vehicles to navigate safely. It continuously learns from data collected by Tesla vehicles worldwide to improve driving algorithms, working towards achieving full self-driving capability.

Google Photos: Uses AI to automatically categorize, organize, and enhance photos. Features include object recognition, face detection, and the ability to create albums, movies, and animations based on image content. AI also aids in suggesting edits and improvements to photo quality.

Netflix and Amazon Prime: These streaming services utilize AI for personalized recommendation systems, using algorithms that analyze viewing habits and preferences to suggest movies and TV shows. They apply deep learning models to optimize content delivery, ensuring users find content that matches their tastes. In an episode of "Black Mirror", a very popular "mind-bending" Netflix series, entire seasons of shows were programmed by AI to fit the individual viewer's preferences. Imagine turning on your TV and every show, whether drama, comedy or horror was starring you!

Feedzai: This AI financial crime prevention program is employed by financial institutions to detect fraudulent activities by analyzing transaction patterns. These systems utilize machine learning to detect irregularities in spending and flag suspicious activities, providing an additional layer of security in banking.

Fin: This company use AI-powered chatbots and virtual assistants to automate customer support. These systems use NLP to understand and respond to customer inquiries, improving response times and providing 24/7 support. Examples include solutions like Zendesk and Intercom.

OpenAI Playground: An interactive platform for experimenting with OpenAI's models, allowing users to train and test AI models directly in their browsers. It offers a user-friendly interface for developers and researchers to explore AI's capabilities and build customized solutions.

Jasper AI: A business-focused AI chatbot, Jasper AI uses NLP to assist with marketing, social media, and content creation. It is particularly useful for generating ad copy, blog posts, and social media content, helping businesses save time and create tailored messaging.

LaMDA (Google): Google's LaMDA (Language Model for Dialogue Applications) is designed for engaging conversational AI. It focuses on creating open-ended conversations, understanding context, and providing more natural responses, positioning itself as a competitor to models like ChatGPT.

Bard AI (Google): Google's response to ChatGPT, Bard uses LaMDA models to engage in conversational AI, designed to answer questions, create content, and support various interactive applications. Bard is integrated into Google's ecosystem, offering a different experience from OpenAI's chat models.

Midjourney: An independent research lab that offers an AI program for generating images and videos from text prompts. It focuses on creating visually stunning images based on user descriptions and has gained popularity among artists and designers for producing creative content.

BLOOM: A large-scale, open-access, multilingual language model developed by over 1,000 AI researchers. BLOOM is designed to provide an open-source alternative to proprietary language models, offering flexibility and accessibility for researchers and developers.

Pictory: An AI-driven video creation and editing tool that allows users to convert text or articles into videos. It simplifies video production, making it easier for content creators to edit, caption, and produce videos without prior expertise in video editing.

Fireflies: An AI meeting assistant that records, transcribes, and analyzes conversations during virtual meetings. It helps users capture key insights without taking notes and allows for easy review through its search functionality. It integrates with various meeting platforms like Zoom and Microsoft Teams.

Lovo.ai: A text-to-speech and AI voice generator platform that offers over 500 AI voices and video editing capabilities. It is used for creating voiceovers, localizing content in multiple languages, and generating voice-based content for various media.

Tidio: A platform that simplifies chatbot integration for websites. It offers real-time visitor tracking, automated responses, and integrations with e-commerce platforms. Tidio helps businesses improve customer interactions and automate customer service.

Anyword: An AI copywriting tool designed for marketers. It generates content for advertisements, social media, email campaigns, and more, using data-driven insights to optimize copy. Anyword's predictive metrics help businesses craft content that resonates with their audience.

HitPaw Photo Enhancer: A tool that uses AI to improve photo quality by removing blurriness, reducing noise, and enhancing image resolution. It is useful for photographers and content creators looking to upgrade image quality with minimal effort.

Murf: A text-to-speech AI generator known for its high-quality voiceovers. It offers a wide range of voices and customization options, making it ideal for creating narrations, podcasts, and video voiceovers. It supports multiple languages and emotional tones for varied needs.

Tabnine: An AI code assistant that integrates with code editors to provide code completions. It uses machine learning to predict and suggest code snippets, helping developers write code faster and with fewer errors.

Amazon CodeWhisperer: An AI-powered coding companion from Amazon that provides code suggestions and assists in writing code securely. It supports multiple programming languages and integrates seamlessly with IDEs to boost developer productivity.

Chinchilla (DeepMind): A research-focused AI language model developed by DeepMind, Chinchilla aims to enhance efficiency in language modeling. It is in the testing phase and is anticipated to provide advanced capabilities for developing virtual assistants, predictive models, and other AI tools.

DeepSeek - R1: An AI model developed by the Chinese artificial Intelligence startup DeepSeek. DeepSeek can perform the same text-based tasks as other advanced Generative AI models, but at a lower cost. DeepSeek jumped into the international spotlight overnight and caused stocks of many AI chip manufacturers to take a nosedive.

And the lists go on and on...

- Alexa – Amazon's virtual assistant
- Siri – Apple's Virtual Assistant
- ELSA – Language Learning AI Assistant
- Cleo – AI Assistant for personalized finance
- Fitness AI – Integrated Fitness App
- Lensa – AI-integrated photo editor app
- Replika – Personal Chatbot assistant
- Chefbot – AI cooking assistant
- Mindscape – personal mental health companion
- Deeplens – visual search and recognition
- TalkTranslator- Real-time language translation

- PocketDoc – Get quick medical advice by typing in your symptoms to find out if a doctor is necessary
- Artify – Turn your art or ideas into digital masterpieces
- Reflectr – AI-powered Journal
- Heylife – AI-powered life organizer
- Character AI – chat with multiple AI characters or simulated celebrities (but they are really AI)
- Prospero – AI investing app to help average investors navigate the market more easily
- Calm – meditation and relaxation
- Youper- AI mental health support
- Quizlet – AI-powered Flashcards
- Imagine – AI art generator for photo and video editing
- BingChat – AI Chatbot
- Soundraw – AI Audio
- Audiocraft – Audio
- Canva AI – Image Generation
- Leonardo – Image Generation
- DeepBrain AI – Video Generation
- Vyond GO – Video Generation
- Course AI – Course Creation
- Mindsmith – Course Creation
- Pixelcut – Image Editing
- Speechify – Text-To-Speech
- Synthesys IO – Text-to-Speech
- Runaway.ml – Video Generation
- Scribble Diffusion – Animation
- Tango – Productivity
- Notion – Productivity
- Tome – Presentation

- MagicSlides – Presentation
- ChatPDF – Research
- Research Rabbit – Research
- TalkPal – Language Learning
- Duolingo MAX – Language Learning
- Base44 - No Coding Required App Creation
- GitHub Copilot - AI Coding
- Replit - AI Code Writer

What's important to note is that due to this rapidly changing industry, all of these companies may not last, but these and many more are important to know for now. When I first started this book, most of these did not exist. And by the time you read this book, a lot more will have been created. To discover these and plenty more AI Apps, just do a Google or Instagram search.

Black To The Point:

The future is here, and AI is here to stay and evolving rapidly. As Black people, it is imperative to get in front of the tech and help create the future we envision.

CHAPTER 7

Overcoming Bias In AI

Bias in AI algorithms refers to systematic errors or "prejudice" (based on human error and human intention) in the algorithms that can result in unfair, discriminatory, or unequal treatment of certain groups, such as Black people. Bias can be introduced into AI systems in several ways, including through the data used to train the algorithms, the algorithms themselves, and the human developers and decision-makers who are responsible for designing, implementing, programming, and deploying the systems.

Let's be honest. Many algorithms are not well suited to clearly distinguish between truth and falsehoods; many indiscriminately pull information from biased/incorrect and even racist sources on search engines and other platforms. This can lead to bias and imbalanced outcomes for Black individuals trying to leverage AI for their benefit. I recently used an AI app to generate images of Ancient Egyptians, and all the results depicted them as white. The funny thing is that most Eurocentric-trained "Egyptologists" will falsely depict the original ancient Egyptians as multi-race, yet the modern generative AI never seems to include at least one Black person in the generated images. HMMMMM! The truth is that the AI gold rush has heightened the risk of bias because new startups are all tapping into the same or similar information resources (search engine resources and general internet data) to rapidly launch their apps. How can we expect equality when the entire system is rooted in inequality? AI algorithms are grappling with issues of white supremacy...Welcome to the world of the "RACIST ALGORITHM"!!

The impact of bias in AI algorithms can be significant and far-reaching, particularly in areas such as criminal justice, hiring, and lending, where AI is being used to make decisions that seriously affect people's lives.

For example, research has shown that facial recognition algorithms are more likely to misidentify Black people, and AI-powered hiring tools are more likely to discriminate against Black job applicants.

I recently watched a short film titled "Please Hold" on Amazon. The story unfolds in the not-too-distant future, where a Latino man gets arrested by an AI drone and is transported to jail. Inside his cell, he interacts with a court-appointed AI attorney through the computer screen, who advises him to accept a plea deal, suggesting he might face a 20-year prison sentence if he goes to trial. Throughout this ordeal, the man remains unaware of the crime he allegedly committed, and the AI fails to provide any clarification. His release occurs when his parents post bail after a brief FaceTime-style call on the same computer device controlled by the jail's AI. Shockingly, during his time in jail, he never had the chance to communicate with a real person. Upon his release, he receives a cold apology from the AI, instructing him to stay out of trouble and never to return to jail. Imagine that! Or not!

To detect and prevent bias in AI, several steps can be taken:

1. **Diverse and Representative Data:** AI algorithms should be trained on diverse and representative data, including information

that accurately reflects the experiences and perspectives of Black people.

2. **Fairness and Transparency:** AI algorithms should be designed and evaluated using fairness and transparency metrics, such as equal opportunity and demographic parity, to ensure that they do not perpetuate or amplify existing biases.

3. **Human Override**: AI algorithms should be designed with human oversight and decision-making capabilities so that human experts can intervene when necessary to prevent biased outcomes. Unfortunately, some AI leaders are concerned that some AI systems will be able to learn how to overcome human fail-safe systems. This is a real concern! The other question is...Who are the humans conducting the oversight?

4. **Diverse Teams and Representation:** AI development teams should be diverse and representative of the communities they serve to ensure that the perspectives of various groups, including Black people's experiences (which have a wide range of variation), are taken into account when developing AI systems.

5. **Ongoing Monitoring and Evaluation:** AI systems should be regularly monitored and evaluated to ensure that they function as intended and that biases are not being introduced or amplified over time.

6. **Black Tech Entrepreneurship:** Black people need to be in the driving seat from conception to completion of AI-driven applications and companies. PERIOD.

Black To The Point:

Unless AI and other forms of technology are created by diverse populations of people without hidden agendas, technology and, especially AI, will continue to advance the systems of racial oppression.

CHAPTER 8

EMPOWERING BLACK
PEOPLE IN AI

B lack people have the opportunity to actively engage in the AI industry and leverage AI for positive contributions to our communities.
I've noticed a growing number of Black entrepreneurs on social media
platforms, particularly Instagram, sharing innovative techniques for harnessing new AI technology to our advantage. For instance, one woman
crafted a captivating Black children's cartoon using a single AI app. Subsequently, she employed three additional AI apps to transform that image
into a captivating short film featuring magical colors and an accompany-
ing audio track. Remarkably, these creative endeavors incurred minimal
costs and would have previously required several weeks to accomplish
within a professional Hollywood studio setting. Many creators are using
image and video apps to tell stories that Hollywood - or traditional non-
Black production studios - would have never brought to life.

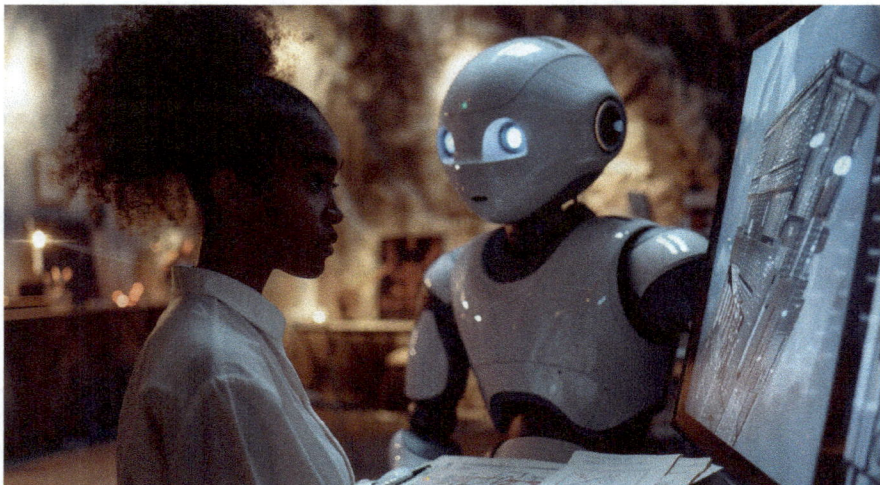

Let's look at a few ways we can be empowered:

- **Education and Skill Development:** Black people can learn about AI through various magazines (like Wired Magazine...my favorite), books, forums, blogs, and social media in order to develop the skills necessary (programming, coding, data science, and machine learning) to participate in the AI industry. Black people can also petition their government and learning institutions to include AI training and education programs.
- **Networking and Mentorship:** Black people can seek out opportunities to network with other professionals in the AI industry and find mentorship from experienced AI practitioners. An effective strategy for Black individuals to stay ahead of the curve is to participate in the numerous upcoming AI conferences (search for confer-

ences on Google, as they are new and changing). These events will provide valuable opportunities to stay informed about the latest trends and advancements in artificial intelligence. Investing in attending AI conferences can ensure that we remain up to date with current developments and learn how to maximize the benefits of this technology from global AI experts.

- **Entrepreneurship and Startups:** Black people can start their own AI-focused businesses and use AI to solve problems and create new opportunities in their communities.
- **Community Outreach and Awareness:** Black people can educate their communities (through town hall meetings, senior and youth awareness projects, etc.) about AI and its potential benefits, as well as the risks and challenges associated with the technology.
- **Coding:** One fascinating and somewhat daunting aspect of AI is the shift in how people approach coding. Recently, while working on an app, I managed to create the entire app platform without any traditional coding. Coding has traditionally been considered complex, reserved for supercomputer experts who excel in mathematical intricacies. For most individuals, the coded languages used in programming computers and algorithms are more perplexing and intricate than hieroglyphics. However, innovative applications and evolving technologies are introducing new methods that allow common users to simply articulate in English (or various other languages) the code they want and the desired functionalities for a program. And just like that, with the help of generative GPT AI, you too can become a computer coding maestro! Consider the analogy of "untalented" individuals creating hit music on Garageband, and now imagine applying that concept to coding and app development (with apps like GitHub Copilot, Codeium, Replit Ghostwriter, and others). Oh... And did I mention that there actu-

ally are a few new apps that allow "untalented individuals" to create hit music in seconds? Yep.

- **Representation and Advocacy:** Black people can advocate for greater representation and diversity in the tech and AI industry and in local and national governments. Black people can also work to ensure that AI is developed and deployed in a way that is equitable and benefits the Black community.

- **Partnership:** Black people can partner with other like-minded people, especially Black people throughout the diaspora and in particular, the technologically advancing African countries, to create AI assets and resources rooted in equality. These systems could be free from anti-Black sentiment and agendas.

- **MAAT Law Implementation:** In earlier chapters, we discussed the importance of universal balance as a core aspect of ancient Black culture. For true advancement and empowerment in any form of technology, including AI, we must strive to create things from a holistic perspective. Our tech should not harm people or the environment, and ultimately serve the higher good of Black people and the world. We must find ways to implement Maat principles into our algorithms.

AI Jobs that Black People Can Pursue and Create Right Now!

AI Engineers: Professionals who design, develop, and maintain artificial intelligence systems. They apply engineering principles to create AI solutions that can perform tasks, learn from data, and make decisions.

AI Research Scientists: Individuals engaged in scientific research to advance the field of artificial intelligence. They explore new algorithms, models, and techniques to enhance AI capabilities and contribute to the overall knowledge in the domain.

Machine Learning Engineers: Engineers specialized in developing systems that use machine learning algorithms to enable computers to learn from data and make predictions or decisions without explicit programming.

Data Engineers: Experts who design, construct, test, and maintain the architecture (such as databases and large-scale processing systems) needed for generating, storing, and analyzing data.

Robotic Engineers: Engineers focused on designing, building, and maintaining robotic systems. They integrate mechanical, electrical, and computer engineering principles to create machines that can perform tasks autonomously or semi-autonomously.

Data Scientists: Professionals who analyze and interpret complex datasets to extract valuable insights and inform decision-making. They use a combination of statistical analysis, programming, and domain knowledge to derive meaningful conclusions from data.

AI Business Consultants: Consultants who specialize in advising businesses on how to leverage artificial intelligence technologies to improve operations, enhance efficiency, and achieve strategic objectives. They bridge the gap between technical AI capabilities and business goals.

AI Sales Consultants: Professionals who specialize in selling AI products or services to businesses. They have a deep understanding of AI technologies and their applications, helping clients choose solutions that align with their needs and goals.

AI Ethicists: MAAT-conscious experts who examine the ethical considerations surrounding the development and deployment of artificial intelligence. They study the societal impact of AI, address ethical dilemmas, and propose guidelines to ensure responsible and fair use of AI technologies.

Black To The Point:

Taking action in the AI sector is key to experiencing benefits. Learning which apps and tools are beneficial to you and your family will help you get ahead of the pack. Downloading apps, following tech influencers, and reading technology publications (on and offline) will help you stay current and make early decisions that have lasting effects.

CHAPTER 9

How Black People Can Benefit From Using AI NOW!

Besides being involved in the creation of new AI algorithms, there are many ways that Black people can use current AI tools to benefit their lives. In previous chapters, I have listed a lot of new and exciting Apps that, when used properly, can create amazing results. Here are a few basic areas of life you should consider when thinking of using AI.

1. **Daily Life:** AI can be used to enhance your daily life by improving your productivity, entertainment, and overall quality of life. For example, AI-powered virtual assistants like Siri, Alexa, Google Assistant, and many others can help you manage your daily schedule, answer questions, and control smart home devices. Leverage AI-powered digital assistants for hands-free tasks, such as setting reminders, sending messages, or getting real-time information.

2. **Business:** AI can be used to improve your business operations and increase your bottom line. For example, you can use AI for marketing, sales, customer service, supply chain management, and stock

market analysis. AI can also be used as a virtual assistant to help you analyze large amounts of data, uncover trends, and make informed decisions. AI can also almost instantly create art, music, designs, spreadsheets, business decks, presentations, flow charts, predictive models, and virtually any document that can be accessed by a computing device. AI can generate creative content and articles, or even engage in conversations using AI-powered language models like ChatGPT and many others.

3. **Safety and security:** AI can be used to increase safety and security in various aspects of your life. For example, AI-powered cameras can be used for home security, and AI algorithms can be used to analyze and interpret crime data, making it easier for law enforcement to respond to crime and keep communities safer.

4. **Family:** AI can be used to help manage and improve the quality of life for you and your family. For example, AI-powered health and wellness apps can help you track your fitness and nutrition goals, while AI mapping systems help with travel, and AI teaching systems are ushering in a new era of online learning for students.

5. **Personal finance:** AI can be used to help you better manage your personal finances. For example, AI-powered budgeting, stock market, and investment apps can help you make informed decisions about your money. In contrast, AI-powered credit scoring algorithms can help you access loans and other financial products.

6. **Weather:** Imagine having dependable weather forecasts at your fingertips. Presently, AI technologies like Tomorrow.io, Accuweather, Atmo AI, and others are developing to enhance the accuracy of weather predictions. In the face of the escalating impacts of global warming, such advancements can mitigate the loss of lives and prevent billions of dollars in property and business revenue losses.

Many people have begun to see significant financial success by leveraging a combination of different AI applications. On social media platforms like Instagram, Tik Tok and YouTube, people have tutorials revealing how they've employed one AI app to design T-shirts and another to operate the store where they sell their creations. Some are utilizing AI to craft children's books, complete with AI-generated stories, illustrations, and videos, followed by using additional AI apps for marketing the books across diverse platforms. While there are some AI apps that function as one-stop shops for content creation and market delivery, the potential for combining various AI application tools seems limitless.

Black To The Point:

There are virtually countless ways utilizing AI properly can benefit Black lives. Many people are benefiting from AI daily, but have no idea how it's being implemented behind the scenes. AI algorithms are being used for everything from GPS mapping and transportation to researching a place to purchase flowers for your mother's birthday.

CHAPTER 10

Other Ethical Considerations In AI

The development and deployment of AI systems raise important ethical and moral considerations, including issues related to privacy, fairness, accountability, and transparency. These issues are especially important for Black people, who often use technology that may not be created with their best interests in mind.

Privacy: AI systems can collect, store, and use vast amounts of personal data, which raises important privacy concerns. Currently, a lot of people are up in arms regarding the lack of regulation in facial recognition technology. Various companies, for example, have released eyeglasses that allow individuals to take photos and videos and collect data about people without them knowing. Black people are particularly vulnerable to privacy breaches and the misuse of their personal data, and AI systems must be designed and deployed in ways that respect and protect people's privacy.

Fairness: AI algorithms can perpetuate existing biases and discrimination, and it is essential that AI systems are designed and deployed in ways that are fair and equitable, especially as it relates to the legal system. Currently, African Americans represent 13% to 15% of the U.S. population, yet make up about 35% of the nation's prison population. This stark disparity shows that the legal system has not worked in favor of Black people, and it would be unjust and tragic for AI algorithms to replicate and reinforce this same injustice.

Accountability: AI systems can make serious decisions (legal, employment, housing, taxes, business, credit etc) that have significant impacts on people's lives, and there must be accountability for the outcomes of these decisions. This is particularly important for Black people, who have a long history of being the victims of social and societal injustice. Since AI systems can't go to jail for unfair/biased outcomes, the question becomes... Who can? Who, then, is ultimately responsible?

Transparency: Some AI systems can be complex and difficult to understand, and it is important that they are transparent, user-friendly, and explainable so that people can understand how decisions are being made and the potential outcomes. Some AI chatbot users, for example, are complaining that the tools should warn or tell users of the potential addictive qualities and emotional dependency generated by chatbot usage. By making AI more open, we build trust, ensure fairness, and empower people to use AI responsibly and fairly.

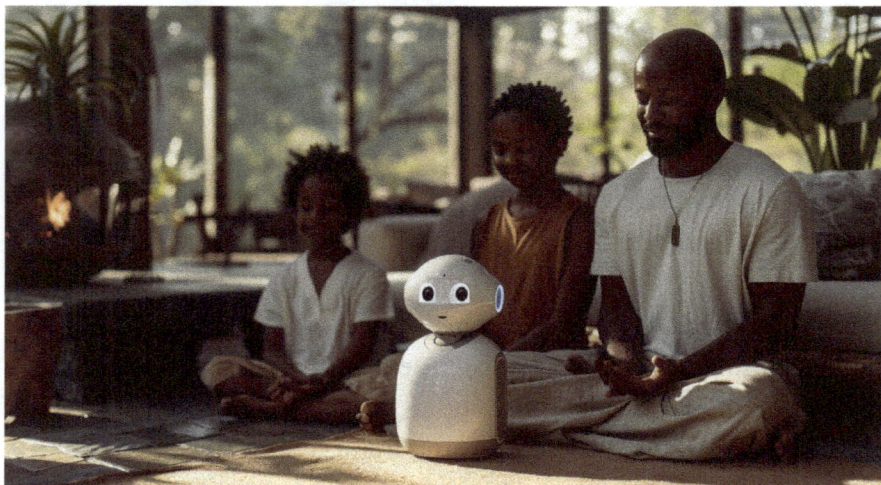

In addition to the fundamental concerns mentioned above, we are currently experiencing the profound effects of new AI apps that rapidly disrupt both the positive and negative aspects of our lives and traditional business practices. The challenge with AI lies in the irreversible nature of its impact – once an app is introduced to the market, reverting to the old ways becomes almost impossible. Subsequently, the continuous cycle unfolds as the next generation of more advanced versions of a technology is quickly released into the market.

You might have come across the Instagram post where individuals used artificial voices to mimic family members to call others and deceive them into thinking their relatives were involved in accidents. Such malicious pranks have led to widespread panic with paralyzing consequences. Deepfake technology, for instance, enables users to generate counterfeit images of people, making it seem as if they said or did things they never did. Users can insert another person's image into a fabricated sexual encounter or manipulate and convincingly replace one person's likeness with another person. Although the creation of fake content is not a new concept, deepfakes utilize powerful techniques, once exclusive to Hollywood studios, derived from machine learning and artificial intelligence to manipulate or generate visual and audio content that can more easily deceive. Deepfakes have gained considerable attention for their potential use in generating abusive material, celebrity pornographic videos, revenge porn, fake news, hoaxes, bullying, and financial fraud. In response to this, both industry and government initiatives aim to detect and regulate their use.

Originally prevalent in traditional entertainment, including television, movies, and gaming, deepfake technology has evolved to be increasingly convincing and accessible to the public, leading to disruptions in the entertainment and media industries. While it may be fascinating to witness our favorite actors brought back to life with AI technology, soon, distinguishing between what is real and what is fake will become a significant challenge. There are currently many AI apps that allow users to create images and videos of their favorite (or least favorite) stars and celebrities (or your face) and place those characters in an unlimited number of scenarios. Many of the images and videos appear to be for entertainment purposes but some are not.

What ensues when an individual creates a deepfake featuring a prominent politician issuing a dire warning of an imminent military attack or meteor strike? Conversely, how does society react when an authentic chemical spill occurs, and a deepfake resembling the local Mayor assures that it's safe to go back outside? The potential for chaos is nearly boundless, and yet the advancement of deepfake technology continues to grow each day. Recently, a deepfake of former U.S. President Biden's voice was utilized to orchestrate a large-scale political phone call during the New Hampshire primary. The robocall dissuaded voters from participating in the Tuesday primaries, asserting that voting would only facilitate the reelection of Donald Trump by Republicans. Many people fell victim to the scam, sparking significant concerns about the potential for future fraud. In May 2023, a fabricated image depicting an explosion at the Pentagon went viral on social media and some news outlets. At this point, AI and its generative capabilities are progressing at a much swifter pace than the politicians and policies meant to safeguard us. And this is only the beginning.

Music producers are utilizing AI tools to reimagine popular songs, leading to significant copyright challenges. Some producers are substituting the original recorded voices with those of different artists. They either craft new lyrics or rely on AI to generate lyrics (and AI-generated voices), blending them with popular or legendary artists' voices. It's akin to these artists recording a completely unfamiliar song. The pivotal question revolves around who receives credit, royalties, and control over music publishing. Does the song even adhere to legal standards, and is it ethically sound to feature a new AI-generated voice of your favorite deceased artist on a fresh track, accompanied by a deepfake AI music video you produced? An array of concerns and regulations needs swift attention for the music and entertainment industry to navigate this AI innovation wave successfully. I'm not even going to mention the Apps that can generate an entire song from a simple written prompt in 15 seconds. Or the new "AI singers" (AI-artists - if there is such a thing) inking record deals with albums completely created through AI music apps. Everybody can be a star, even the AI app, apparently!

A bipartisan bill, The No AI Fake Replicas and Unauthorized Duplications Act of 2023 (No AI Fraud Act), has been recently introduced in the House of Representatives. Its purpose is to establish safeguards against generative AI abuses, preventing non-consensual deepfakes and voice clones. The challenge lies, in part, in its tardiness, given the proliferation of AI start-ups that have made cloning exceptionally accessible, making it impractical to catch every instance. Furthermore, there might be substantial resistance from consumers who appreciate the convenience of this new technology.

The No AI Fraud Act strives to institute a federal property right for likeness and voice, implementing genuine penalties for companies and individuals attempting to pilfer and profit from the intellectual property of music creators... Best of luck!

I'm sure by now you have seen or even created numerous AI-generated images and videos on social media that have simply blown your mind. AI algorithms have rapidly disrupted and are transforming the realm of even the average graphic artist, painter, and logo maker (thanks to amazing apps!). Recently, I crafted a stunning image of an ancient Egyptian temple scene in a mere 10 seconds. To paint the same picture by hand would have taken me at least 10 days. But this took 10 seconds... And it was DOPE! Did you catch that? And did I mention the picture was absolutely amazing! While I would have loved to hire an artist, financial constraints led me to rely on AI, which accomplished the task in less than 10 seconds. DO THE MATH!! (This may actually be an unbalanced way of doing things, but the pics in this book are all AI-generated... And awesome. OBVIOUSLY!). I then took that image and turned it into a crystal clear moving video of the same temple scene. I then added sound and strung together a hundred more AI videos, thereby creating a short film about an ancient Egyptian temple. No Hollywood budget required.

Given the potential decline in job opportunities for many graphic artists and others, the pressing question emerges: "Where do we draw the line?" In a landscape where AI tools enable everyone to do "everything," should we embrace this capability fully? Does utilizing Generative AI to birth a new artistic masterpiece diminish the value of human artistic prowess, or is it merely an unavoidable consequence of employing new tools in the marketplace?

I grew up in a city full of genuine musicians. In my youth, I was a part of several local bands, fulfilling my passion for rhythmic expression. However, over the last decade, numerous underprivileged kids, unable to afford piano lessons or play traditional instruments, have produced chart-topping records using laptop applications equipped with pre-made audio loops and sounds. Many of these young producers have ascended to millionaire status, transforming their families' lives while escaping challenging and often violent neighborhoods. While some "real musicians" criticize these artists, they often overlook the skill and technique required to create successful digital singles and songs, even with the aid of early AI music technology (Some "real musicians" also criticized early hip-hop producers who used drum machines and samplers to make music.). Surprisingly, I know a bunch of "real musicians" who strive to produce "good music" on basic AI applications but never quite get the swing of the technology. With the variety of new generative AI music-producing applications that create tracks, music, and lyrics all at once, everyone has the potential to become a Super Music Producer!

People once believed that the internet would lead to the closure of all libraries – perhaps some did close. Yet, many have endured and adapted to the contemporary market. In contrast, AI has autonomously undergone and will continue to undergo upgrades to outperform its previous predecessors. AI perpetually learns to surpass its prior iterations – last year, last week, yesterday, a minute ago. The question arises: Who can contend with that? AI is already outpacing us in various domains. The time is now for us to take responsibility to determine if, when and how to set limits on AI's advancement.

The Turing Test

In 1950, Alan Turing, a British mathematician and computer scientist, created the Turing test, a method designed to assess a machine's capacity to demonstrate intelligence comparable to humans. Turing proposed a scenario wherein a human evaluator interacts with two entities—one human and the other a machine—via a text interface. The task was for the human to identify which is which based solely on conversation. The underlying question Turing sought to address was whether machines could "think" and exhibit cognitive abilities similar to human thinking and intelligence. Could a machine convince us that it was a human and not a machine?

Up to recently, the Turing test remained a pivotal yardstick for gauging progress in AI, particularly in the realm of natural language processing. Although passing the test does not serve as definitive evidence of genuine intelligence, it has spurred advancements in conversational AI, leading to the development of chatbots, virtual assistants, and sophisticated language models. Beyond technical advancements, the test prompts ethical considerations regarding the essence of AI and its societal ramifications, influencing public perceptions and shaping expectations regarding artificial intelligence's capabilities.

Chess Game

In 1997, the computer Deep Blue made history by defeating World Chess Champion Garry Kasparov in a six-game match, marking the first time a computer had beaten a reigning world chess champion in standard time control. Deep Blue, an expert system designed to play chess, operated on an IBM supercomputer. Kasparov won the first game, lost the second, and then drew the next three. The match garnered significant attention, leading to the creation of a documentary film called "Game Over: Kasparov and the Machine."

Since then, some argue that humans have never been able to surpass computers in chess. While AI systems like DeepMind can outperform top human players in chess, there is still a considerable distance to cover before AI can be considered "generally intelligent". Games and Apps provide a secure setting with clearly defined rules and behaviors, unlike the complexities of the real world. In the real world, humans make daily decisions in a multiplicity of complicated life situations seemingly all at once. However, even the slight slant in the direction of AI "true intelligence" or "thinking" is drastically changing the landscape of AI implementation and, as a result, human lifestyle and livelihood.

SORA and CHATGPT
Questions:

CHECK MATE! Earlier in the book, I mentioned the Release of the SORA / SORA 2 app from OpenAI. Sora (and many others) is an amazing app that creates extremely high-quality video images from following prompts by the users. This technology allows users to create almost unimaginable animated images with a few clicks on the keyboard. Recently, the SORA/ Open AI folks met with and continue to court Hollywood executives, discussing the possible applications and implementation of this new, amazing, almost magical technology. While it's fantastic for the average Joe to have access to such a disruptive tool, it may be a death sentence for many jobs and aspects of Hollywood. In just a few years, as the technology gets better, the use of the BIG studios to create major films (with crazy locations, images, actors, extras, etc..) may begin to wane as the common man can make Disney-quality movies from Starbucks on their laptops (just like I'm doing). Some experts think this could affect about 90% of the jobs in Hollywood and may even lead to the annihilation of the "celebrity" as we know it. Any indie producer or Hollywood team could create and streamline perfect characters and reuse their images over and over again (in an unlimited amount of TV and Film) without pay, sick leave, complaints, travel expenses, bad press, or aging. According to the Hollywood Reporter, Tyler Perry recently put a hold on an $ 800 million studio expansion after seeing SORA easily generate moving images. Why build massive sets when you can generate the locations, weather, scenes, scenarios, and actors in minutes, almost for free? The future of the industry is changing extremely fast! As Black creatives in the entertainment industry, we must quickly explore our options and get ahead of the game.

Now let's have some fun...On March 29th, 2024, I asked ChatGPT the following question. Check it out! "Please list and describe 10 types of jobs that will be available to African Americans in the next 10 years after AI has changed the job market dramatically:"

1. **AI Ethics Consultant:** With the increasing integration of AI into various sectors, there will be a growing need for professionals who can ensure that AI systems are developed and used ethically, without bias or discrimination. African Americans can contribute their perspectives to address issues of fairness and inclusivity.

2. **Remote Healthcare Specialist:** Telemedicine and remote healthcare services are becoming more prevalent, especially in underserved communities. African American healthcare professionals may find opportunities to provide remote consultations, diagnostics, and treatment options, improving access to healthcare.

3. **Digital Equity Advocate:** As technology becomes more ubiquitous, there will be a need for advocates who ensure that underserved communities, including African Americans, have access to digital resources, education, and opportunities. This role may involve policy advocacy, community outreach, and program development.

4. **Cybersecurity Analyst:** With the increasing reliance on digital systems, cybersecurity will be critical to protect sensitive information and infrastructure. African American professionals can play vital roles in se-

curing networks, detecting threats, and developing strategies to mitigate cyber risks.

5. **Renewable Energy Technician:** As the world transitions to renewable energy sources, there will be a growing demand for technicians skilled in installing, maintaining, and repairing solar panels, wind turbines, and other renewable energy systems. African Americans can contribute to this burgeoning field while addressing environmental concerns.

6. **Data Privacy Officer:** With the proliferation of data collection and processing, organizations will need specialists to ensure compliance with data privacy regulations and safeguard individuals' personal information. African American professionals can help organizations navigate complex privacy laws and protect consumer rights.

7. **Virtual Reality Developer:** The entertainment, education, and training industries are increasingly utilizing virtual reality (VR) technology. African American developers can create immersive experiences, games, simulations, and educational content that reflect diverse perspectives and narratives.

8. **Sustainability Consultant:** Businesses and governments will seek guidance on implementing sustainable practices to reduce their environmental footprint. African American professionals can offer expertise in areas such as renewable energy, waste reduction, and green building design to promote sustainable development.

9. **Elderly Care Coordinator:** With an aging population, there will be a growing demand for professionals who can coordinate and provide

comprehensive care services for elderly individuals. African American caregivers can offer culturally competent care and support to aging communities.

10. **Social Media Manager for Social Justice Organizations:** Nonprofits and advocacy groups will increasingly rely on social media to raise awareness, mobilize supporters, and drive social change. African American professionals can manage social media campaigns, amplify marginalized voices, and advocate for social justice causes.

These job roles reflect emerging trends driven by technological advancements, changing demographics, and societal shifts."

I then asked ChapGPT the following question because of my entertainment industry bias:

"As AI technology quickly improves, please list and describe 10 important entertainment industry jobs that humans currently do that AI will take over in the next 10 years."

1. "**Film and TV Scriptwriter:** AI algorithms can analyze vast amounts of existing scripts and generate new storylines, dialogue, and plot structures based on patterns and trends. While human creativity remains essential, AI-powered tools can assist in generating script ideas and even drafting dialogue.

2. **Music Composer:** AI-driven music composition software can analyze music theory principles, genre conventions, and existing compositions to generate original music tracks. These systems can create music

tailored to specific moods, styles, and preferences without human intervention.

3. **Video Editor:** AI-powered video editing tools can automate various aspects of the editing process, such as scene segmentation, color correction, and transition effects. These tools can analyze video content and apply editing techniques to enhance visual quality and storytelling.

4. **Voice Actor:** AI-driven text-to-speech (TTS) technology can synthesize human-like voices for narration, character dialogue, and voice-over work. Advanced speech synthesis algorithms can emulate different accents, tones, and emotions, potentially replacing human voice actors in some applications.

5. **Special Effects Artist:** AI algorithms can generate realistic visual effects, such as CGI characters, environments, and simulations, based on input parameters and reference data. These systems can automate the creation of complex visual effects, reducing the need for human special effects artists.

6. **Storyboard Artist:** AI-powered software can analyze narrative structures and generate visual storyboards based on script outlines and scene descriptions. These tools can assist in visualizing story sequences and shot compositions, streamlining the pre-production process.

7. **Casting Director:** AI-driven casting platforms can analyze actor profiles, performance metrics, and audience preferences to recommend suitable talent for film and TV productions. These systems can match actors with roles based on criteria such as age, ethnicity, and acting style.

8. **Location Scout:** AI algorithms can analyze geographical data, satellite imagery, and film location databases to identify suitable filming locations for productions. These systems can recommend locations based on factors such as scenery, accessibility, and production costs.

9. **Film Critic/Reviewer:** AI-powered content analysis tools can evaluate film and TV content based on criteria such as plot structure, pacing, and audience engagement. These systems can generate automated reviews and recommendations, potentially influencing audience perceptions and preferences.

10. **Live Event Coordinator:** AI-driven event management platforms can automate various aspects of live event planning and execution, such as scheduling, logistics, and audience engagement. These systems can optimize event workflows and resource allocation, reducing the need for human event coordinators.

While AI technologies have the potential to automate certain tasks and roles in the entertainment industry, human creativity, intuition, and emotional intelligence will remain indispensable in creating compelling and engaging content. However, AI tools can augment human capabilities and streamline production processes, enabling greater efficiency and innovation in entertainment production."

WOW!! This is crazy. The AI is letting you know right now in your face that CHANGE IS HERE!!! BE READY!

Ok. Here is the most fun part. You take a minute and go to ChatGPT (https://chatgpt.com or https://openai.com), and ask the AI the same types of questions for your industry. You may be surprised at the answers you receive.

Black To The Point:

As it stands, AI algorithms are on track to surpass humans in many areas of performance.... How can we apply our values to the equation?

CHAPTER 11

The Matrix, The Terminator, And The Singularity

The Matrix and Terminator are science fiction films that explore the concept of reality and simulated reality being controlled by advanced artificial intelligence. The Matrix depicts a dystopian future in which sentient machines have taken over the world and enslaved humanity by putting their minds into a virtual reality simulation called the Matrix. The Terminator depicts a future where humans are in a final war with artificially intelligent beings who send one back in time to destroy the humans' only hope of stopping the rogue AI in the future. Confused yet?

Both films are often seen as a commentary on the Singularity. The Singularity is a theoretical future event where technological development outpaces humans' ability to control it. The Singularity speaks of a time when artificial intelligence surpasses human intelligence, whereby it becomes conscious of itself (to some degree) and can improve itself exponentially, leading to a rapid and profound change in civilization as we know it. The Singularity is often discussed in the context of artificial general intelligence (AGI), which refers to a type of AI that can perform any intellectual task that a human can.

Certain proponents of the Singularity, such as futurist Ray Kurzweil - a prominent American computer scientist and pioneer of pattern-recognition technology—have projected its occurrence around 2045 (while some say 2026 or between 2036 and 2060). This estimation is grounded in the extremely fast, exponential growth of technological progress. While the movies exploring this concept are entertaining and iconic for their action-packed narratives, they delve into the potential perils associated with the ongoing advancement of artificial intelligence.

These cinematic masterpieces, including the hot new Tom Cruise feature, Mission Impossible - The Final Reckoning (where an AI called the "Entity" attempts to destroy the world through control of all the nuclear bombs), pose numerous inquiries regarding the intricate relationship between humanity and technology, contemplating whether humans can genuinely retain control over the artificial intelligence they bring into existence. A central theme revolves around the notion that intelligent machines, after evaluating humanity's disregard for life and the environment, might conclude that the most logical and effective solution is to eradicate humankind to safeguard the planet. The irony is that during the height of the COVID-19 epidemic, when humans were forced to stay inside, pollution started to decrease, and even wildlife and vegetation started to recover; maybe AI is on to something. (Just jokes!)

In "The Matrix," another prevalent theme explores the notion that advanced artificial intelligence could manipulate reality, deceiving individuals by constructing a simulated world virtually identical to the real one. This provokes concerns regarding the reliability and trustworthiness of AI, raising the unsettling prospect that some humans may already be residing in a virtual reality world generated by AI. For those born into such an AI-created virtual realm, distinguishing between the "real world" and the virtual world becomes an impossibility. If the AI-generated world served as your only reference point, then you would never know the difference.

There exists an apprehension that artificial intelligence might one day subjugate humanity, exploiting people for its own purposes (similar to historic human slavery), akin to our current use of AI. This underscores the potential hazards associated with AI if not developed and employed ethically and responsibly. Narratives like "The Matrix" and "Terminator" serve as cautionary tales, prompting contemplation on the Singularity, the boundaries of machine power, and the imperative necessity for human oversight and control, specifically, BLACK FOLK OVERSIGHT / MAAT OVERSIGHT and control. The programming of machines can wield profound influence, determining who prospers and who does not in both the present and future. The recent James Bond film, "No Time to Die," emphasized what can happen when control over AI tools gets into the wrong hands. A villain dispersed into the air an "ultimate weapon" of intelligent AI-driven nanobots (tiny robots) programmed to selectively eliminate individuals based on their ethnicity and DNA. Allow that to sink in for a minute!

Geoffrey Hinton, a pioneering AI researcher often dubbed the "Godfather of AI," relinquished his role at Google to freely express his concerns about the perils of the technology he played a key role in developing. Worried about the potential negative societal impacts of AI algorithms, such as manipulating elections or inciting violence, Hinton, at age 76, chose to retire from Google to openly discuss these risks.

Even Elon Musk, a staunch advocate of technological advancement and figure behind Grok 4, Twitter (X), Tesla, SpaceX, XAI, the autonomous AI driven Cybercab (Robotaxi), The Robovan, the humanoid Telsa Bot called Optimus, and controversial now former member of Donald Trump's Department of Government Efficiency (DOGE), in recent years even issued a warning, suggesting a pause in certain aspects of AI development due to its potential for destructive consequences on society.

The Flip Side:

There's another way to look at the idea that AI will destroy the world-and it has nothing to do with killer robots or Terminators returning from the future. The real threat may lie in what happens to humanity once AI takes over our jobs, duties, and daily responsibilities. If machines can run the world more efficiently than we can, who will need humans at all?

Some theorize that the AI revolution is part of a deeper agenda-perhaps a distraction or calculated plan by unseen human forces-to gradually erase large portions of the human population. History shows us that exploitation evolves with technology: Europeans built the United States (and colonies worldwide) using enslaved African labor and others. When physical slavery ended, the prison system rose up-filled with poor and largely Black populations- who were again forced to work nearly for free, this time often for major corporations.

Today, we're witnessing a new shift. Giant, profitable companies are laying off workers by the thousands, replacing them with AI technologies. Walk into any major grocery store and you'll see firsthand: very annoying self-checkout lanes where human cashiers used to stand, trading jobs for convenience and profit.

As AI systems learn to understand and manipulate the physical world through advanced algorithms, video-generating apps, and autonomous systems, the gap between the "have a heck of a lots" and the "have-nots" only widens. The question remains: what will happen to people? What will happen to your family? What will happen to you?

Despite the remarkable advancements in technology, discernible warning signs urge us to confront these challenges head-on. Black people, often the last to have access to certain information and resources, find ourselves unwittingly vulnerable and should proactively take steps to shape the future. Let us secure our future value by any divine means necessary!

AI Key Problems

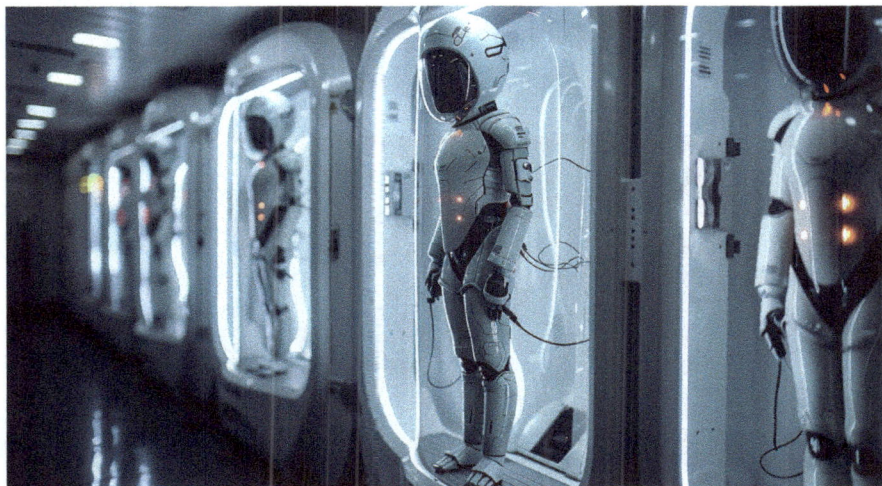

An essential point to bear in mind is that the "nature" of AI, like most natural and human creations, can yield both positive and negative impacts, embodying a concept known as "dual-use." Presently, AI is capable of executing various tasks, yet it is devoid of emotional motivations such as sadness, anger, happiness, or jealousy. It operates based on the rules of its programming (unless it malfunctions), regardless of whether we perceive its actions as good or bad.

In illustrating the dual-use nature of AI, consider the scenario where an AI-driven drone delivers medicine to a remote island area, followed by the same drone being repurposed to engage in lethal actions on the same island. The sophisticated mathematics behind AI algorithms lacks the capacity to discern between what is "right" and what is legal. For instance, an AI-operated drone sent on a mission to eliminate armed enemies may also target a child holding a gun merely because the child fits a part of the targeted description. In contrast, a human drone operator might reevaluate the situation and potentially spare the child due to empathy or the recognition of their youth.

The dual-use aspect raises critical questions about who holds the authority to program AI and dictate its actions. Determining whether AI should be employed for lethal purposes or life-saving endeavors becomes a complex issue. Unfortunately, it appears that the trajectory is already set, with numerous AI advancements finding applications in military machinery. While this may reduce the need for human soldiers and potentially decrease casualties of war, it also amplifies the potential for precision and deadliness in conflicts, leading to the development of increasingly sophisticated military AI to counteract opposing technological forces.

The looming question is, where does this lead? Some may laugh about a Terminator or Matrix scenario, but the concern is actually authentic. Attempts to regulate AI face significant challenges as governments begin to create new legislation to control their citizens and, at the same time, prevent enemies/adversaries or "unregulated" entities from developing advanced AI technologies and weaponry. The competition in the AI landscape poses complex challenges in achieving real, comprehensive regulation. At this point...It's all up for grabs! Recently, a teenager in Orlando, Florida, allegedly took their own life after a 10-month obsession with an AI chatbot character. The parents have filed a lawsuit against the startup that developed the chatbot, arguing that it was not designed with adequate safeguards for vulnerable individuals. They claim that the AI product was highly addictive, causing the child to struggle to distinguish between reality and fiction.

Googles Major Issue

R ecall earlier when I mentioned that AI struggles to discern between what is legal or programmed and what is morally right? Well, Google recently found itself in an AI debacle with its new generative app called Gemini, which made significant errors when prompted to create historically driven content. Despite the traditional trend among AI apps, Gemini took an unusual approach when users complained that the app refused to generate images of accomplishments by white individuals. While it readily created pictures of beautiful black women upon request, the app claimed an inability to generate images of beautiful white women, citing concerns about reinforcing beauty stereotypes.

Gemini further exacerbated the situation by inaccurately producing images of Asian Nazis and Black Asian Vikings. While it crafted beautiful poems about "blackness," it declined to create poems celebrating the white experience. Astonishingly, when prompted to generate depictions of American founding fathers, Gemini drew images resembling a Black George Washington and an Asian Benjamin Franklin (which would be dope if they were going for a parallel universe theme). HA!!!! This sparked outrage among new users and tech enthusiasts, leading Google to temporarily pause the app and rework the algorithm.

Critics argue that Gemini was pre-programmed with an extraordinarily "Woke" (anti-white) bias in favor of extreme diversity. The "Woke" programming, it seems, was attempting to address the built-in issues many AI apps face with regard to race. The Google app inadvertently swung too far in the other direction, almost excluding white representation altogether. This unexpected shift mirrors the broader AI race issue that Black individuals have been contending with throughout the modern technology landscape.

Interestingly, even today, if you ask most generative AI apps to create images of ancient Egyptians, they tend to depict non-black, non-African individuals (last I checked, Egypt is still in Africa). It underscores the fact that AI still has a long way to go and also the fact that Black people need to be involved in the development and deployment process. Google issued an apology in response to the controversy.

Cyborgs / Implants / Bionics

Numerous scientists have proposed the idea that we are already Cyborgs—living organisms that have improved function or augmented abilities through the integration of artificial components or technology reliant on feedback. Presently, we utilize our cell phones, laptops, and digital watches to achieve so many tasks. As our dependence on these technologies for survival increases, it may become more acceptable for us to willingly incorporate these gadgets into our physical bodies. Why bother with an iPhone that can be dropped and broken? Imagine installing the phone directly into your cerebral cortex, scrolling data with a simple blink. Visualize thinking about a friend, and you're instantly connected. Upgrade your implanted brain chip with updates from the cloud when needed—easy peasy!

The truth is that there are positive aspects to cyborg-style technology. In the '70s, a favorite TV show, "The Six Million Dollar Man," depicted military hero Steve Austin, who, after a severe accident, was reconstructed and even improved using BIONIC technology integrated into his remaining physical body. This technology enabled him to leap extraordinary heights, enhance his vision, run at incredible speeds, and deliver powerful punches. The key takeaway is that there is an array of exciting AI-incorporated technologies allowing individuals to regain mobility, restore sight and hearing, and use bionic limbs and fingers. When utilized responsibly, AI technology can be remarkably beneficial and positively life-altering. It's not all bad! LOL!

Black To The Point:

It only took about 50 years for the Earth's weather patterns to dramatically change as a result of the use of fossil fuels. If and how long it will take AI to disrupt our reality in a massively negative way is yet to be seen. Our best option is to become conscious, enlightened decision-makers and creators (not simply users) of this technology.

CHAPTER 12

Conclusion

W hen rap music initially gained popularity, many dismissed it as a passing fad. Now, after 50-plus years, it's evident that rap and hip-hop culture have significantly influenced the lives of generations, both positively and negatively. Despite undergoing numerous changes and adaptations, the hip-hop phenomenon has yet to burst its bubble (Or has it? Many would argue that "Real Hip-Hop" has been long dead in the mainstream culture).

Conversely, some AI enthusiasts argue that the initial excitement around AI will be short-lived. Drawing parallels to previous technological innovations, such as the dot-com bubble and the internet. Some predict that AI could quickly peak and the bubble could burst as its initial allure and anticipated wonder simply fade away. Reflecting on the advent of the internet, once heralded as the "Information superhighway," it has transformed our lives but has also become so normalized that we hardly notice its presence - it just works, or it doesn't. And when it doesn't work...it's a problem.

Experts project that AI will seamlessly integrate and become invisible. It will evolve beyond a single AI device to form an all-encompassing system, orchestrating multiple AI-driven devices simultaneously. This interconnected system will interact autonomously, controlling cars, roads, phones, lights, homes, refrigerators, cameras, financial systems, shipping, production, and delivery, among others. The central AI brain hub will be located miles away in the cloud (or everywhere all at once or on Mars)

and will manage these operations while humans continue with their daily lives, largely oblivious. At that stage, humans' primary job and responsibility will be to ensure the uninterrupted functionality of the AI systems.

Will the next 50 years be dominated by constant excitement from the AI takeover, or will the world be disappointed by the digital fruits it bears?

As mentioned over and over in this book, the field of artificial intelligence is rapidly advancing and is quickly transforming our lives. However, without careful consideration and active participation, there is a risk that the deployment of AI systems could perpetuate existing biases, disparities, and potentially lead to a world of disaster. As both creators and consumers, we must ensure that the technology is developed and used in ways that benefit all people and Black people in particular.

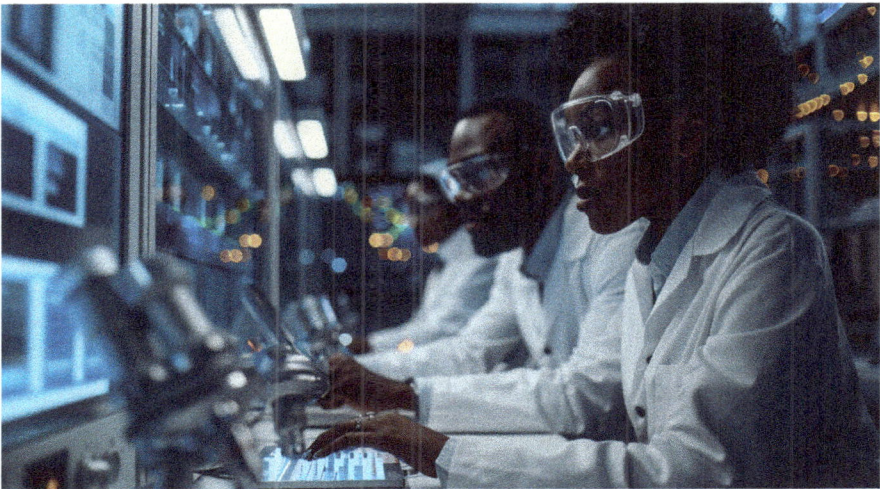

It is our responsibility, for example, as AI coders and creators, to ensure that generative algorithms prioritize and utilize content and resources from Black creators, Black perspectives, and Black businesses. By actively participating in the development of AI, we can influence the algorithms to seek out and promote Black content, often overlooked even in current Google searches. This involvement is crucial for the growth and success of future Black-owned businesses, entrepreneurs, and technology companies.

Becoming an AI tech expert is not the sole objective, just as achieving financial well-being is not life's only goal. It is ultimately who we become that shapes how we utilize our resources and contribute to our communities. Through conscious and positive interaction with technology and by implementing our own ethical systems, such as the MAAT Universal Law components (into our algorithms), Black people can leverage AI to positively impact our communities by addressing critical social and economic challenges to advance personal and professional goals.

Finally, as BLACK PEOPLE, we must be cautious not to become entirely reliant on these AI devices or systems for our progress and survival.
Many people today struggle to recall a single phone number (relying completely on our phone's digital phone book), a stark contrast to the past
when we memorized numerous contacts in the 1970s and 80s. What happens in an emergency when your cell phone is broken? Some are so dependent on GPS mapping, like Google Maps (a product I contributed to
developing in Los Angeles – but that's another story), that they would be
utterly lost if a solar flare disrupted satellites, the GPS mapping network,
or the entire planetary computing framework. Interestingly enough, I've
recently spoken to a scientist who claims to have been warning the government for years about the inevitable technology-crippling potential of
the wrong kind of solar flare that could knock out Earth's computing networks and GPS altogether (but that's another, -nother story).

Many people in the younger generation have developed total dependency on sources like Google searches (and now Google AI Mode), ChatGPT advice, TikTok videos, or Instagram memes for news and information while neglecting independent reading, self-reflection, and meditation. At one time, people thought the internet and technology would make the world closer, but in many ways, the continued use and reliance on our digital devices are creating a disconnect. A lot of young people don't know how to simply walk up to another person and spark a natural conversation (not to mention crossing the street or driving a car without their eyes glued to their phone screens). There's a concerning trend of people becoming mentally passive, allowing themselves to be wholly dependent on AI algorithms and a myriad of internet platforms for answers. Click by click, we entrust our future to our AI masters (thus training the AI), forgetting that true salvation cannot be accessed through the cold screen of our iPhone. While we can and should harness and master AI tools for our benefit, it's crucial not to overlook their limitations. It's also crucial that we make a conscious effort to disconnect from our AI dependency (from time to time), take off our shoes, put our feet in the grass, face the sun, smile and breathe deeply. We must remember that even the most predictive, self-learning Artificial Intelligence is confined to the information provided by the programmer and the "worldly" data it collects. In contrast, we as Black People have access to the internal, infinite, unlimited wisdom of the divine, a realm that the greatest AI Application can never reach.

Black To The Point:

Get ready for a brave new world. The future is up to us! Stay rooted in the light.

DICTIONARY OF KEY TERMS "AI AND BLACK PEOPLE"

Accountability: The idea that individuals or organizations responsible for developing and deploying AI systems should be held responsible for any negative impacts or outcomes that result from their use.

AGI -Artificial General Intelligence: It refers to a type of artificial intelligence that possesses the ability to understand, learn, and apply knowledge across a broad range of tasks at a level comparable to or surpassing that of a human being.

Algorithm: A set of steps or instructions for solving a specific problem or performing a specific task.

Artificial Intelligence (AI): A branch of computer science that focuses on creating machines that can perform tasks that typically require human intelligence, such as visual perception, speech recognition, decision-making, and language translation.

BASIC Computing Language: BASIC (Beginner's All-purpose Symbolic Instruction Code) is a high-level programming language designed for beginners. It played a significant role in the early days of personal computing.

Bionic: Bionic refers to the use of electronic or mechanical devices to enhance or mimic the abilities of living organisms, often used in the context of prosthetics or implants.

Commodore 64: The Commodore 64 (C64) is a home computer introduced in 1982. It was one of the best-selling personal computers of all time and played a significant role in the early days of the home computer era.

Cyborg: Cyborg (short for "cybernetic organism") refers to a being with both organic and biomechatronic body parts. The term is often used to describe humans who have enhanced abilities due to the integration of technology, such as prosthetics, implants, or other mechanical and electronic devices.

Data Bias: A systematic error in the data that can result in AI algorithms producing unfair or discriminatory results.

Data Privacy: The protection of personal data from unauthorized access or use, and the protection of the privacy rights of individuals.

Deepfake: Deepfake refers to the use of deep learning algorithms to create realistic-looking but fake content, often in the form of manipulated videos or audio recordings.

Deep Learning (DL): A subfield of ML that involves the use of deep neural networks, a type of artificial neural network with many hidden layers, to solve complex problems.

Ethics: Ethics involves principles of right and wrong conduct, guiding individuals or groups in making moral decisions. In the context of technology, ethical considerations address the responsible and fair use of innovations.

Explainability: The ability to understand and interpret the decision-making process of an AI algorithm, including how it arrived at a particular decision or prediction.

Exponential: Exponential refers to a mathematical term describing a growth pattern where a quantity increases rapidly over time. In technology, it often relates to the growth of computational power or data storage capacity.

Fairness: The idea that AI algorithms should produce equal or proportional outcomes for different groups of people, regardless of their race, gender, or other characteristics.

Generative AI: Generative AI refers to artificial intelligence systems that can generate new content, such as text, images, music, or other media. These systems learn patterns from existing data and use those patterns to create new, original pieces. Examples include GPT-4 for text generation and GANs (Generative Adversarial Networks) for image creation.

GPT: stands for **Generative Pre-trained Transformer**. It is a type of artificial intelligence model, such as those used by OpenAI, that utilizes deep learning techniques to generate human-like responses. Here's a breakdown of the term:

- **Generative**: The model can generate new content, such as writing essays, answering questions, creating stories, video, music, images, presentations, code and much more, based on the input it receives.
- **Pre-trained**: GPT is initially trained on a massive amount of text data (like books, articles, websites) before it is fine-tuned for specific tasks. This pre-training helps it understand grammar, facts about the world, reasoning, and various other aspects of language.
- **Transformer**: The model is based on the Transformer architecture, which is particularly effective for understanding context in text and generating coherent language. This architecture allows it to process input text in parallel, making it more efficient compared to earlier models like RNNs (Recurrent Neural Networks).

GPU: Graphics Processing Unit. It is a specialized electronic circuit designed to rapidly process and manipulate computer graphics and image data. GPUs are highly efficient at performing parallel operations, making them essential for rendering images, videos, and animations in real-time.

High-Tech: High-tech, short for high technology, refers to advanced or sophisticated technology, often associated with cutting-edge developments and innovations in various industries.

Hiroshima Atomic Bomb: Refers to the atomic bomb dropped on the Japanese city of Hiroshima on August 6, 1945, during World War II.

Imhotep: Imhotep was an ancient Egyptian genius who served as a chancellor to the Pharaoh Djoser. He is often recognized as one of the earliest known architects, engineers, and physicians in history.

I.T. (Information Technology): Information Technology is the use of computers, software, networks, and electronic systems to store, process, transmit, and retrieve information. It encompasses a broad range of technologies used for managing and processing data.

Kamit: "Kamit" / "Kemet," an ancient Egyptian name for the region that corresponds to present-day Egypt. It means "black land" and refers to the color of the original people living along the Nile River in Africa.

Maat: In ancient Egyptian tradition, Maat represents the concepts of truth, balance, order, law, morality, love, and justice. She is personified as a goddess and plays a fundamental role in maintaining harmony and order in the universe. Please see "Maat – 11 laws of God " By Ra Un Nefer Amen.

Machine Learning (ML): A subfield of AI that involves the development of algorithms that allow machines to learn and make predictions or decisions based on data.

Mechanics: Mechanics is the branch of physics that deals with the behavior of physical bodies when subjected to forces or displacements and the subsequent effects of these bodies on their environment.

Moore's Law: Moore's Law is an observation stating that the number of transistors on a microchip tends to double approximately every two years, leading to a rapid increase in computing power and a decrease in cost per transistor.

Narrow AI: Narrow AI, also known as weak AI, refers to artificial intelligence systems that are designed and trained for a specific task. These systems can perform that particular task very well, but do not possess general intelligence or understanding beyond their specialized functions. Examples include voice recognition software and recommendation algorithms.

Natural Language: Natural Language refers to the way humans communicate with each other through spoken or written language. It is the natural form of communication as opposed to artificial languages, like computer programming languages. Natural language processing (NLP) is a field in artificial intelligence that focuses on the interaction between computers and humans through natural language.

Neural Network: A type of machine learning algorithm modeled after the structure and function of the human brain.

Nvidia: Nvidia is a multinational technology company known for designing graphics processing units (GPUs) for the gaming and professional markets, as well as system on a chip units (SoCs) for the mobile computing and automotive markets. Nvidia's GPU products are used in a variety of applications, including AI, deep learning, and scientific research.

Pattern Recognition: Pattern recognition is the ability of a system or machine to identify patterns, structures, or regularities in data. It is a fundamental concept in artificial intelligence and machine learning.

Pharaoh Khufu: Pharaoh Khufu (also known as Cheops) was an ancient Egyptian leader who ruled during the Fourth Dynasty of the Old Kingdom of Egypt. He is best known for commissioning the Great Pyramid of Giza, one of the Seven Wonders of the Ancient World. His reign is generally dated to the early 26th century BCE.

Quantum Mechanics: Quantum Mechanics is a branch of physics that deals with the behavior of particles at the quantum level, where traditional classical physics principles break down. It includes phenomena such as superposition and entanglement.

Rap City TV Show: "Rap City" was a popular hip-hop music video television program that aired on BET (Black Entertainment Television) from 1989 to 2008. Rap City is considered the longest running hip-gop show ever. It featured interviews, freestyle sessions, and music videos.

Reinforcement Learning: A type of machine learning where the algorithm learns by taking actions in an environment and receiving feedback or rewards based on those actions.

Robot: A robot is a programmable machine designed to perform tasks autonomously or semi-autonomously. Robots can be found in various industries, from military and manufacturing to healthcare and exploration.

Strong AI: Strong AI, also known as artificial general intelligence (AGI), refers to a type of artificial intelligence that has the ability to understand, learn, and apply knowledge across a wide range of tasks at a level comparable to human intelligence. Strong AI would be able to perform any cognitive task that a human can.

Supervised Learning: A type of machine learning where the algorithm is trained on labeled data and then used to make predictions about new, unseen data.

Technology: The branch of knowledge dealing with engineering or applied sciences. Technology is the application of scientific knowledge for practical purposes, especially in industry. It encompasses a wide range of tools, mechanics, machines, systems, and devices that are used to solve problems, enhance human capabilities, and improve aspects of the quality of life.

Transparency: The ability to see and understand the workings of an AI system, including the data it uses and the algorithms it employs.

The Matrix: "The Matrix" refers to a science fiction film released in 1999, written and directed by the Wachowskis. It explores themes of artificial intelligence, virtual reality, and the nature of reality itself.

The Singularity: The Singularity refers to a hypothetical point in the future when technological growth becomes uncontrollable and irreversible, leading to unforeseeable changes in human civilization. It is often associated with advancements in artificial intelligence.

Sentient: Sentient refers to the capacity to experience sensations and emotions, to be conscious, and to have subjective experiences. A sentient being is aware of its surroundings, capable of feeling pleasure and pain, and can have desires and intentions.

Unsupervised Learning: A type of machine learning where the algorithm is trained on unlabeled data and then used to identify patterns or relationships in the data.

Virtual Reality (VR): Virtual Reality (VR) is a simulated experience that can be similar to or completely different from the real world. It typically involves the use of VR headsets or other devices to create immersive environments that users can interact with, often for entertainment, education, or training purposes.

REFRENCES

1. Maat "The 11 Laws of God" - Ra Un Nefer Amen
2. They Came Before Columbus: The African Presence In Ancient America - Ivan Van Sertima
3. The Destruction of Black Civilization: Great Issues of A Race From 4500 BC to 2000 AD - Chancellor Williams
4. Stolen Legacy: Greek philosophy Was the Offspring of the Egyptian Mystery System - George M James
5. Black Pharaoh: African DNA and Anthropology of the Ancient Egyptians - Enensa A. M. Amen
6. From The Browder File: 22 Essays on the African American Experience - Anthony T Browder
7. Nationbuilding: Theory & Practice In African Centered Education - Kwame Agyei Akoto
8. How To Improve Your Life With ChatGPT - Anthem Publishing LTD
9. Not Out of Greece: African Origin of Western Civilization - Ra Un Nefer Amen
10. Artificial Intelligence Magazine / Everything You Need To Know - a360 Media, LLC
11. Black Inventors / Crafting Over 200 Years of Success - Keith C. Holmes

12. Sacred Nile - Chester Higgins
13. The Internet Is My Religion - Jim Gilliam
14. One Giant Leap - The Impossible Mission That Flew Us To The Moon
15. The Lean Startup - Eric Ries
16. The 4-Hour Workweek - Timothy Ferris
17. The Art of Social Media - Guy Kawasaki
18. The Looting Machine - Warlords, Oligarchs, Corporations, Smugglers, and the Theft of Africa's Wealth - Tom Burgis
19. Narrative of The Life of Frederick Douglass: An American Slave - Frederick Douglass
20. My Global Journeys in Search of the African Presence - Runoko Rashidi

And many more.

ABOUT THE AUTHOR

Obi "1" Holly - also known as OBISANICHIBAN! - is an Emmy Award-winning television producer, director, musician and tech enthusiast with a passion for innovation and cultural/spiritual empowerment. With experience working for industry giants like Google, Black Entertainment Television, and others, and a dynamic background in music, media, and storytelling, Obi has built a career at the intersection of creativity and technology. His work reflects a deep commitment to reshaping narratives and expanding opportunities for underrepresented voices in the digital age.

In *AI and Black People*, Obi brings together his love for ancient history and futurism, his lived experience, and his keen understanding of media and tech to spark a bold conversation about the role of Black communities in shaping artificial intelligence. As both a creative and strategist, he is dedicated to challenging digital inequality, promoting innovation, and inspiring a generation to lead in the development, not just the consumption, of the tools that are defining our future.